ELEMENTARY
PROBABILITY
THEORY

ELEMENTARY PROBABILITY THEORY

MELVIN HAUSNER

Washington Square and University College
New York University

SPRINGER SCIENCE+BUSINESS MEDIA, LLC

Library of Congress Cataloging in Publication Data

Hausner, Melvin, 1928-
  Elementary probability theory.

  "A Plenum/Rosetta edition."
  Reprint of the ed. published by Harper & Row, New York, in series: Harper's series
in modern mathematics.
  1. Probabilities. I. Title.
[QA273.H4 1977]   519.2   77-813

  ISBN 978-1-4613-5706-3   ISBN 978-1-4615-1753-5 (eBook)
  DOI 10.1007/978-1-4615-1753-5

First paperback printing 1977. This edition is published by arrangement with
Harper & Row, Publishers, Inc. *Elementary Probability Theory* was first published as
a volume in Harper's Series in Modern Mathematics.

Copyright © 1971 by Melvin Hausner
Originally published by Plenum Publishing Corporation in 1971

PREFACE

This text contains ample material for a one term precalculus introduction to probability theory. It can be used by itself as an elementary introduction to probability, or as the probability half of a one-year probability–statistics course. Although the development of the subject is rigorous, experimental motivation is maintained throughout the text. Also, statistical and practical applications are given throughout.

The core of the text consists of the unstarred sections, most of chapters 1–3 and 5–7. Included are finite probability spaces, combinatorics, set theory, independence and conditional probability, random variables, Chebyshev's theorem, the law of large numbers, the binomial distribution, the normal distribution and the normal approximation to the binomial distribution. The starred sections include limiting and infinite processes, a mathematical discussion of symmetry, and game theory. These sections are indicated with an *, and are optional and sometimes more difficult.

I have, in most places throughout the text, given decimal equivalents to fractional answers. Thus, while the mathematician finds the answer $p = 17/143$ satisfactory, the scientist is best appeased by the decimal approximation $p = 0.119$. A decimal answer gives a ready way of finding the correct order of magnitude and of comparing probabilities. Also, in applications, decimal answers are often mandatory. Still, since $17/143$ is only equal to 0.119 to three places, one is confronted with the problem of notation. Should one write $17/143 \approx 0.119$, or $17/143 = 0.119-$, or $17/143 = 0.119$ (three places)? This author simply wrote $17/143 = 0.119$. The reader must therefore be prepared for the convention (well established in common usage) that an equation involving decimals is, in most cases, only an approximation.

I wish to acknowledge my debt to some of the people who helped in the production of this text. Mrs. Eleanor Figer did the typing of the final manuscript. E. Sherry Miller, my student, worked the problems and grew to appreciate the value of the slide rule. Leonard Hausner, my assistant and son, performed the probability experiments. Finally, thanks is hereby given to my class A63.0004 (Introduction to Probability) for submitting to the first dittoed version of the text and for criticizing and correcting much of it.

CONTENTS

Preface   vii
CHAPTER   1.   THE FOUNDATIONS   1
          1.   Experimental Basis for Probability   1
          2.   Mathematical Formulation of Probability   11
          3.   Events   16
          4.   Some Examples   23
          5.   Some Generalizations   30

CHAPTER   2.   COUNTING   37
          1.   Product Sets   37
          2.   Multiplication Principle   45
          3.   Permutations and Combinations   51
          4.   Binomial Coefficients   61
          5.   Applications of the Multiplication and Division Principles   70
         *6.   Cards   75

CHAPTER   3.   GENERAL THEORY OF FINITE PROBABILITY SPACES   81
          1.   Unions, Intersections, and Complements   82
          2.   Conditional Probability   90
          3.   Product Rule   97
          4.   Independence   104
         *5.   Construction of Sample Spaces   112

*CHAPTER   4.   MISCELLANEOUS TOPICS   123
         *1.   Repeated Trials   123
         *2.   Infinite Processes   129
         *3.   Coincidences   137
         *4.   Symmetry   146

CHAPTER   5.   RANDOM VARIABLES   153
          1.   Sigma Notation   154

2. Random Variables and Their Distributions   *158*
3. Expectation   *165*
4. Algebra of Expectations   *175*
5. Conditional Expectations   *180*
6. Joint Distributions   *185*
*7. Introduction to Game Theory   *194*
*8. Symmetry in Random Variables   *200*

CHAPTER   6. STANDARD DEVIATION   *207*
1. Variance and Standard Deviation   *208*
2. Chebyshev's Theorem   *218*
3. Variance of Sums   *222*
4. Independent Trials   *226*
5. The Law of Large Numbers   *231*

CHAPTER   7. SOME STANDARD DISTRIBUTIONS   *235*
1. Continuous Distributions   *235*
2. Normal Distribution   *242*
3. Binomial Distribution   *249*
4. Normal Approximation   *256*
5. Statistical Applications   *262*
*6. Poisson Distribution   *269*

APPENDIXES
A. Values of e^{-x} *276*
B. Values of e^{x} *278*
C. Square Roots   *280*
D. Areas Under the Normal Curve   *281*
E. Binomial Distribution   *282*
F. Poisson Distribution   *285*

Answers   *289*
Index   *307*

*Optional Material.

ELEMENTARY
PROBABILITY
THEORY

CHAPTER 1 THE FOUNDATIONS

INTRODUCTION

Probability theory is a branch of mathematics concerned with the measurement and understanding of uncertainty. Historically, this theory came into being to analyze certain games of chance. But it is clear that uncertainty occurs not only in gambling situations but all around us. When we ask what the maximum temperature will be in Chicago next July 17, or how many traffic fatalities will occur in New Jersey on the Memorial Day weekend, most people will agree (before the event occurs) that there is an element of uncertainty to the answer. An astonishing feature of probability theory is that it is possible to have such a theory at all. Yet this theory is not only possible, it is also one of the most interesting and fruitful theories of pure and applied mathematics.

In this chapter we lay the foundations for the subject by considering the experimental basis and meaning of probability and then formulating the mathematical description. This sets the stage for the more sophisticated theory which will be taken up in the succeeding chapters.

1 EXPERIMENTAL BASIS FOR PROBABILITY

In a world of uncertainty, we soon learn that some things are more uncertain than others. The sex of a child about to be born is usually uncertain, but very likely an expectant mother will not have twins, and almost surely she will not be a mother of quadruplets. The driver of an automobile knows that most likely he will not be involved in an accident, but he is probably aware that an accident can occur, so he possibly will fasten his seat belt. If a pack of cards is well shuffled, it is uncertain whether the top card is a black card, it is unlikely that it is an ace of spades, and it is extremely unlikely that the top 5 cards are the 10, J, Q, K, and ace, all in spades.[1] The possibilities in our world range from impossible, to unlikely, to maybe, to a good bet, to certainty.

These varieties of uncertainty suggest that we *measure* how certain an event is. This can be done by performing an *experiment* many times and observing the results. Let us consider an example.

1 Example

Three dice[2] are tossed. What is the highest number that turns up?

An appropriate experiment is to toss 3 dice and record the highest number that appears. This experiment was actually performed 100 times and the results appear in Table 1.1. The results suggest that 6 is more likely to occur

1.1 Results of the 3-Dice Experiment

Highest number showing	1	2	3	4	5	6
Number of times this high number occurred	1	6	5	12	33	43

as high number than 5. Similarly, 5 appears more likely than 4, etc. The fact that 2 appeared more often than 3 seems to indicate that 2 is more likely as high number than 3. But the table as a whole, and perhaps a healthy intuition, indicates otherwise, and we rather expect that 3 is more likely than 2. This particular experiment was repeated another 100 times with the results recorded in Table 1.2.

1.2 Additional Results of the 3-Dice Experiment

Highest number showing	1	2	3	4	5	6
Number of times this high number occurred	1	3	10	19	26	41

1 In this book we take for granted a familiarity with a standard pack of 52 cards. This pack contains 4 suits—hearts, diamonds, clubs, and spades; the first 2 are red and the last 2 are black. There are 13 cards in each suit: ace (A), king (K), queen (Q), jack (J), 10, 9, 8, 7, 6, 5, 4, 3, and 2.

2 A die is a cube whose 6 faces are engraved with the numbers 1, 2, 3, 4, 5, and 6, respectively. The plural of "die" is "dice."

When we compare Tables 1.1 and 1.2, we should be more surprised by the similarities than the obvious differences. It is as if some benevolent god were watching the experiments and arranged to make the results similar. Later we shall learn how to compute the *probability* that 6 is the highest number showing of the 3 dice that were thrown. This probability turns out to be .421, or 42.1 percent. According to probability theory this percentage will very likely be close[3] to the actual percentage of the time that 6 appears as high number, if the experiment is repeated a large number of times. In fact, this figure is seen to be strikingly close to the mark in Tables 1.1 and 1.2. Similarly, probabilities can be found for a 5 high, 4 high, etc. These probabilities and the percentage results of Tables 1.1 and 1.2 are listed in Table 1.3. In later chapters we shall learn how to compute these probabilities.

1.3   Summary of the 3-Dice Experiment

High number	1	2	3	4	5	6
Percentage of occurrence (first 100 trials)	1	6	5	12	33	43
Percentage of occurrence (second 100 trials)	1	3	10	19	26	41
Probability (percent) (theoretically derived)	.5	3.2	8.8	17.1	28.2	42.1

This example is typical of many probability experiments. Before proceeding to some other examples, it will be useful to state some of the features common to probability experiments.

1. In contrast to many deterministic scientific experiments in which there can only be 1 outcome, there were several possible outcomes of this experiment. These may be conveniently labeled s_1, s_2, s_3, s_4, s_5, and s_6. Here s_5, for example, stands for "the highest number showing on the 3 dice is 5." Similar definitions apply to s_1, s_2, etc.

2. The experiment was repeated many times. Here, we can see why such gambling devices as dice, cards, etc., are useful tools in the experimental probabilist's hands. With comparative ease it was possible to repeat our experiment many times. Clearly, if we wish to find the probability that Mrs. Jones's next baby will be a male, it seems unreasonable (as well as impossible) to try the experiment 100 times! When we call for experimental probabilities, it is required to repeat an experiment many times and to make

3 To be clarified in Chapter 6.

sure that the same experiment is run. We must be sure to shake the dice, or to shuffle the cards well. In the case of Mrs. Jones, we prefer to ask the question, "What is the sex of a child born in Englewood Hospital?" We then apply these results to the unknown sex of Mrs. Jones's unborn baby. Here hospital statistics furnish figures analogous to those in Table 1.1. This "experiment" is repeated many times during a year at the hospital.

When an experiment is repeated many times, it is not the number of times each outcome occurs that will be of primary significance for us. It is the relative frequency or fraction of the time that each outcome occurs that will be most important. Hence we make the following definition.

2 Definition

Suppose that the outcomes of an experiment may be $s_1, s_2, \ldots, s_k$. Suppose this experiment is repeated N times and that the outcome s_1 occurs n_1 times, $\ldots, s_k$ occurs n_k times. Then n_1 is called the *frequency* of s_1 (in that particular run of N experiments), and the *relative frequency* f_1 of s_1 is defined by the formula

$$f_1 = \frac{n_1}{N}$$

with similar definitions for the frequency and relative frequency of each of the other outcomes $s_2, \ldots, s_k$.

Remark. Clearly $f_1, f_2, \ldots, f_k$ may vary from experiment to experiment. For example, in Table 1.1, $k = 6$, and our outcomes are s_1, s_2, s_3, s_4, s_5, and s_6. $N = 1 + 6 + 5 + 12 + 33 + 43 = 100$. Also, $n_1 = 1$, $n_2 = 6$, $n_3 = 5$, $n_4 = 12$, $n_5 = 33$, and $n_6 = 43$. Thus $f_1 = n_1/N = \frac{1}{100} = .01 = 1$ percent, $f_2 = n_2/N = \frac{6}{100} = .06 = 6$ percent, etc. The relative frequencies in Table 1.2 are different, and in any comparison of the two tables it would be necessary to use different notations. Thus we might use $g_1, g_2, \ldots, g_6$ for relative frequencies, and $m_1, m_2, \ldots, m_6$ for the frequencies.

3 Theorem

Suppose $s_1, s_2, \ldots, s_k$ are the possible outcomes of an experiment. If the experiment is repeated several times and these outcomes occur with relative frequencies $f_1, \ldots, f_k$, respectively, then we have

$$0 \leqslant f_1 \leqslant 1, 0 \leqslant f_2 \leqslant 1, \ldots, 0 \leqslant f_k \leqslant 1 \tag{1.1}$$

and

$$f_1 + f_2 + \cdots + f_k = 1 \tag{1.2}$$

This theorem simply states that the percentages are not negative or over 100 percent, and that they add up to 100 percent. The proof is as follows.

We know that s_i occurs n_i times $(i = 1, \ldots, k)$.[4] Thus

$$N = n_1 + n_2 + \cdots + n_k \tag{1.3}$$

We have $0 \leq n_i$, because an event cannot occur a negative number of times. Therefore,

$$0 \leq n_i \leq n_1 + n_2 + \cdots + n_k = N$$

Dividing by N, we have

$$0 \leq \frac{n_i}{N} \leq 1$$

or

$$0 \leq f_i \leq 1$$

Furthermore,

$$f_1 + \cdots + f_k = \frac{n_1}{N} + \frac{n_2}{N} + \cdots + \frac{n_k}{N}$$

$$= \frac{n_1 + n_2 + \cdots + n_k}{N} = \frac{N}{N} = 1$$

This proves the result.

We can now give the *statistical definition* of probability. Suppose an experiment has the possible outcomes $s_1, \ldots, s_k$. These outcomes are said to have the probabilities $p_1, \ldots, p_k$, respectively, if, when the experiment is repeated N times and N is very large, the relative frequencies $f_1, \ldots, f_k$ of the outcomes $s_1, \ldots, s_k$ will in all likelihood be very close to $p_1, \ldots, p_k$. Briefly, $f_1 \approx p_1, \ldots, f_k \approx p_k$ when N is large. (The symbol "$\approx$" is read "is nearly equal to.") This "definition" can be validly criticized on several points. It is vague, because the terms "very large" and "very close" are used. Also, it hedges a bit with the phrase "in all likelihood." Nevertheless, the intuitive idea is clear: If the experiment is repeated many times, the relative frequency f_i of s_i will be close to p_i. Thus, despite the unpredictable outcome of an experiment, the relative frequency of an outcome can be *approximately forecast* if a large number of experiments is to be performed. The reader will note (cf. Table 1.3) that in the dice experiment with $N = 100$, the relative frequencies of the various outcomes never differed by more than 5.1 percent (.051) from the probabilities.

The ability to predict, with high accuracy, the relative frequency of an occurrence that occurs at random is quite astonishing and is probably not fully accepted by the average person. Still, every year we are presented with predictions of perhaps 500 or so traffic fatalities for the New Year's weekend. The outcomes, alas, are always depressingly near the predictions.

We have arrived at a quantitative measure of uncertainty. It is therefore

---

4 This is a brief way of stating: "s_1 occurs n_1 times, s_2 occurs n_2 times, $\ldots$, s_k occurs n_k times." The price we pay for brevity is the introduction of the new letter i.

natural to define an outcome s_1 to be *more likely* than the outcome s_2 if $p_1 > p_2$. (Here p_i is the statistical probability of s_i.) Similarly, we may define *equally likely* events s_1 and s_2 as events with equal probability. We can say that s_1 is *twice as likely to occur* as s_2 if $p_1 = 2p_2$. This means that if a large number of experiments is performed, the relative frequency of s_1 will be approximately twice the relative frequency of s_2. The *range of uncertainty* from impossible to certainty is now expressed by the inequality $0 \leqslant p_1 \leqslant 1$. Here 0 represents impossibility, 1 certainty, and each number in this range represents a degree of uncertainty. (See Fig. 1.4.)

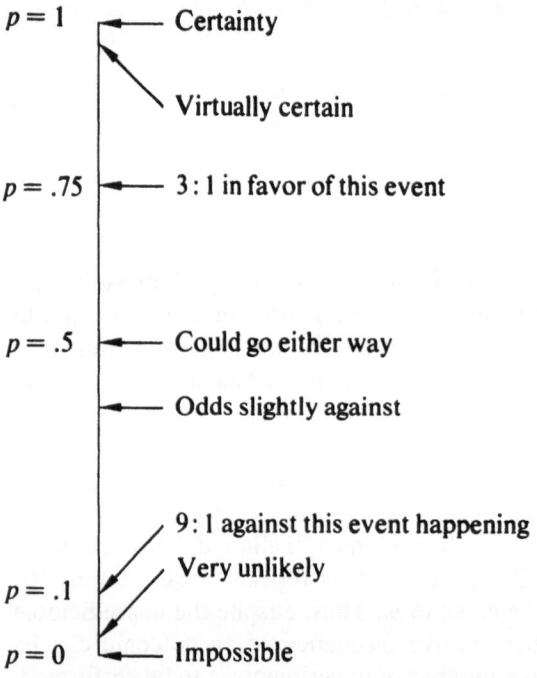

$p = 1$ — Certainty

Virtually certain

$p = .75$ — 3 : 1 in favor of this event

$p = .5$ — Could go either way

— Odds slightly against

9 : 1 against this event happening

Very unlikely

$p = .1$

$p = 0$ — Impossible

1.4 Quantitative Versus Qualitative Descriptions of the Likelihood of an Event

We note here that the so-called "law of averages" is often misapplied, because *absolute* rather than *relative* frequencies are incorrectly used. If an outcome (say tossing a coin and landing heads) has probability .5, we do *not* say that if the experiment is repeated (say) 10,000 times it will very likely occur 5,000 times. For even if it occurred 5,063 times, the *relative* frequency is .5063, and this number is very near .5000, although the actual number of occurrences is 63 more than 5,000.

We now give the results of a few other experiments, together with the relative frequencies of the various outcomes and the probabilities.

4 Example

A coin is tossed. Does it land heads or tails?

The experiment was repeated 500 times with the results shown in Table 1.5. In this example we have 2 outcomes, H (heads) and T (tails), and, correspondingly, 2 frequencies, $n_h = 269$ and $n_t = 231$. Thus $N = n_h + n_t = 500$ (the total number of experiments). The relative frequency of heads is $f_h = n_h/N = 269/500 = .538$, and similarly $f_t = .462$ (see Definition 2). Note that $f_h + f_t = 1$, as in Theorem 3.

1.5 Coin-Tossing Experiment

	Heads	Tails
Frequency	269	231
Relative frequency	.538	.462
Probability	.5	.5

5 Example

A die is tossed until a 6 turns up. How many throws are required?

Experimental results are tabulated in Table 1.6. Here $N = 60$. There are several points of interest here. First, there were infinitely many possible outcomes of this experiment. It was conceivable that the number of throws necessary before a 6 turned up would be 1, 2, 3, etc., indefinitely. (It was even conceivable that a 6 might never turn up.) Because we are mainly concerned in this book with experiments having finitely many outcomes, we merely lumped all outcomes that needed 13 or more throws into *one* outcome—"over 12." Thus we only had 13 outcomes $s_1, s_2, \ldots, s_{12}, s_{13}$, where s_{13} was the "over 12" outcome.

Another point to observe is that in this particular series of 60 experiments, the fluctuations appear to be rather pronounced. The *probabilities* indicate that s_1 is the most likely outcome, with s_2 next likely, etc. Yet we had 6 occurrences of s_{10} and no occurrences of s_9. Also, s_4 occurred most often. These things happen, and when they do, there are at least 4 explanations:

1. There were not enough trials. Thus $N = 60$ is not so large.
2. A very unlikely event happened. (Loosely speaking, we expect occasional miracles.)
3. The dice were not shaken enough to ensure that the tosses were completely random.
4. The results merely *look* unlikely, because the observer is not very sophisticated.

1.6   Waiting for a 6 to Show Up

Number of throws required to obtain a 6	1	2	3	4	5	6	7	8	9	10	11	12	Over 12
Frequency	9	7	7	12	4	4	1	2	0	6	0	4	4
Relative frequency	.150	.117	.117	.200	.067	.067	.017	.033	.000	.100	.000	.067	.067
Probability	.167	.139	.116	.096	.080	.067	.056	.047	.039	.032	.027	.022	.112

Surprisingly, item 4 is often the best explanation for an apparent "miracle." Note, however, that even the other explanations do not negate the idea of statistical probability. We must be willing to imagine a very large series of experiments and we must often concede that an actual series of experiments gives merely a slight indication of what is meant by statistical probability.

EXERCISES

1. An experiment with possible outcomes A, B, or C is repeated several times with frequencies as follows:

Outcomes	A	B	C
Frequencies	119	203	278

Compute the relative frequencies f_A, f_B, and f_C.

2. When 2 dice are tossed, the sum of the numbers that turn up can be any integer between 2 and 12 inclusive. This experiment was performed by the author's assistant with the results indicated. Compute the relative frequencies and compare with the indicated probabilities.

Sum on 2 dice	2	3	4	5	6	7	8	9	10	11	12
Probabilities	.028	.056	.083	.111	.139	.167	.139	.111	.083	.056	.028
Frequencies	6	12	22	32	40	43	31	24	15	18	7

3. (*Births — male or female?*) According to the *Statistical Abstract of the United States, 1962*, statistics for male (M) and female (F) births in the United States during the years 1945–1949 are as indicated in Table 1.7. Compute the year-by-year relative frequencies f_M and f_F. Note that in these tables N is quite large. What conclusions can you draw?

1.7 Live Births (in Thousands), by Sex

	Males	Females		Males	Females
1945	1,467	1,391	1948	1,866	1,771
1946	1,754	1,657	1949	1,872	1,777
1947	1,960	1,857			

Exercises 4 through 9 are essentially experiments to be performed by the reader. In a classroom situation, a large number of experiments can be performed if each student does relatively few experiments. The experimental results should be saved to compare with the probabilities that will be determined later in the book.

4. A pack of cards is well shuffled. Three cards are taken from the top of the deck. How many black cards are there among these 3 cards?

a. State the possible outcomes.

b. Perform the appropriate experiment 50 times and record your results.

c. Compute the relative frequencies. [*Hint:* In this exercise there are clearly 4 outcomes s_0 (no blacks), s_1 (1 black), s_2 (2 blacks), and s_3 (3 blacks). A very good way of keeping count is indicated in the following hypothetical, partially completed table:

s_0	//
s_1	𝍸 /
s_2	𝍸 //
s_3	///

A stroke of the pencil indicates that an event has occurred; every fifth stroke is horizontal.

5. Three dice are tossed. What is the highest number that turns up?

a. State the possible outcomes.

b. Perform the experiment 100 times and record your results.

c. Compute the relative frequencies and compare with the probabilities in Table 1.3.

6. Five dice are tossed. What is the middle number that turns up? (For example, if the dice thrown are, in increasing order, 1, 2, 2, 5, 6, the middle number is 2.)

a. State the possible outcomes.

b. Perform the experiment 100 times and record your results.

c. Compute the relative frequencies.

7. Ten coins are tossed. How many heads turn up? As in Exercises 4 through 6, state the possible outcomes, perform the experiment 50 times, and compute the relative frequencies.

8. The results of Exercise 7 may be used to find the relative frequency with which a tossed coin lands heads. Explain how this may be done. Find this relative frequency using the experimental results obtained in Exercise 7.

9. A coin is tossed until it lands tails. How many tosses will it take? As in Exercise 4, state the possible outcomes, perform the experiment 80 times, and compute the relative frequencies.

Exercises 10 *through* 14 *are discussion problems. There is not necessarily a correct answer.*

10. In the 3-dice experiment (Example 1), is there any reason you would expect a 6 high to be more likely than a 5 high? Similarly, why might you expect (before any experiments are performed) that a 5 high is more likely than a 4 high, etc.?

11. In the coin-tossing experiment of Example 4, is there any reason to expect (before the experiment) that heads is as likely as, or more likely than, tails?

12. In the waiting-for-6 experiment of Example 5, is there any reason to suppose that it is more likely that a 6 will turn up on toss 1 rather than on toss 2?

13. A person tossed a coin 10,000 times. He claims that the first 5,001 tosses were heads and the next 4,999 tosses were tails. When it is objected that this seems very unlikely, he claims that the relative frequency of heads is .5001, which is extremely close to the probability .5. Is the objection valid? Why?

14. In Exercise 6 is there any reason to suppose that the middle number 1 is less likely than 2? That 1 and 6 are equally likely? That 2 and 5 are equally likely?

2 MATHEMATICAL FORMULATION OF PROBABILITY

The injection of mathematics into the study of a physical phenomenon usually has the effect of enormously aiding in the understanding of that phenomenon. Probably the most familiar example is geometry. From its crude, tentative beginnings involving the measurement of length, area, and volume, the brilliant structure of Euclidean geometry was constructed. This structure has proved so successful that the physical applications are now more or less considered to be a rather trivial by-product of it. On the other hand, some topics have resisted a mathematical formulation. (For example, the mathematical study of esthetics has been attempted, but with no apparent success.) The test of any mathematical treatment of a physical phenomenon is its ability to enhance the understanding of the phenomenon.

What is a mathematical formulation? Basically, it reduces the subject to a few simple concepts, which are then studied with the help of definitions, theorems, and proofs. Hopefully, the mathematics then helps to develop one's intuition about the subject in question, while the physical intuition helps to develop and motivate the mathematics. With these preliminary remarks, let us return to probability theory.

In the examples of Section 1, each experiment had several possible *outcomes* $s_1, \ldots, s_k$. Furthermore, each outcome s_i had a certain *probability* p_i, which was thought of as the long-range relative frequency of the outcome s_i. Since the relative frequencies were all between 0 and 1 inclusive, and since they added up to 1 (Equations 1.1 and 1.2), it is reasonable to suppose that the same is true for the probabilities. In the mathematical formulation, the outcomes will be called *elementary events*. The set of all possible outcomes will be called the *probability space S*. In this formulation, we simply think of S as a finite set of elements $s_1, s_2, \ldots, s_k$, where the nature of these elements is irrelevant.

6 Definition

A *probability space S* is a finite set of elements $s_1, s_2, \ldots, s_k$, called *elementary events*. Corresponding to each elementary event s_i is a number p_i called the *probability* of s_i. The numbers p_i must satisfy the conditions

$$0 \leqslant p_i \leqslant 1 \qquad (i = 1, 2, \ldots, k) \tag{1.4}$$

and

$$p_1 + p_2 + \cdots + p_k = 1 \tag{1.5}$$

The terms *sample point, simple event*, or *atomic event* are sometimes used instead of "elementary event." The term "probability space" is used only when each elementary event is assumed to have a probability. If we wish to speak of the set S without regard to probabilities, we call S the *sample space*. Thus the term "probability space" implies that probabilities of elementary events are defined, whereas this is not so for the term "sample space."

In the general theory of probability, a sample space can be an infinite set and Definition 6 would be the definition of a *finite* probability space. However, we do not define this more general concept, and for simplicity we use the term "probability space" instead of the more correct term.

Note that nothing in Definition 6 is said about the numbers p_i being "long-range relative frequencies," but it will do no harm if the reader thinks of p_i in this way. Similarly, there is no harm in geometry when one thinks of a point as a dot on a paper, despite the understanding that a point is an undefined term.

We shall also use the notation $p(s_i)$ (read "probability of s_i" or simply "p of s_i"). Thus $p(s_i) = p_i$. This notation (called *functional notation*) makes clear the dependence of the probability p_i on the elementary event s_i. We shall also let s designate a typical elementary event and $p(s)$ its probability.

Definition 6 is a purely formal one. It permits the construction of probability spaces with relative ease. For example, we may take $S = \{A, B, C\}$, with $p(A) = .8, p(B) = .1, p(C) = .1$, and we have constructed a probability space. (Note that the numbers .8, .1, and .1 satisfy Equations 1.4 and 1.5) This probability space may be put into tabular form as in Table 1.8. The fact that this example is so simple merely shows that the idea of a probability space is a simple one. Of course, this example is arbitrary and sterile, too. We are still in the elementary stage of the subject.

1.8   Probability Space Illustrated

s	A	B	C
$p(s)$	.8	.1	.1

We may now summarize the correspondence thus far obtained between physical reality and its mathematical formulation. Physically, an experiment is performed with several possible outcomes. Each of these outcomes is thought to have a statistical probability. These probabilities satisfy Equations 1.1 and 1.2. We abstract this situation into a mathematical formulation as follows. A set S, called a probability space, is given. Each element s of S is called an elementary event and determines a number $p(s)$, called the probability of s. These probabilities satisfy Equations 1.4 and 1.5. The correspondence between reality (the experimental situation) and the mathematical model (the probability space) is given in the following table:

Experimental situation	Mathematical formulation
Possible results of an experiment	↔ Elementary events s of a probability space S
Possible outcome s	↔ Elementary event s
Statistical probability of s	↔ $p(s)$

In future sections we shall attempt to formulate mathematical concepts in such a way as to reflect experimental situations. In this way the program outlined in the second paragraph of this section can be carried out.

It is considered respectable to mix the two languages. In the simple coin-tossing experiment (Example 4), the outcomes were heads or tails. A probability space might be $S = \{H, T\}$, with $p(H) = .5, p(T) = .5$. It would be considered quite pedantic in everyday use to speak of "the probability of the elementary event H." We are permitted to speak instead of "the probability of tossing a head," and we shall occasionally make similar simplifications throughout the text. The context usually makes it clear whether it is the statistical probability or the more abstract probability which is under consideration.

Definition 6 does not teach us *how* to compute probabilities; it merely presents these probabilities. In a practical situation, where probabilities are unknown, it is necessary to have more information to find these probabilities. One of the simplest assumptions which leads to an immediate and explicit value for the probabilities is the assumption that the various sample points have equal probability. In the real world we would say that the various outcomes are equally likely.

7 Definition

A probability space $S = \{s_1, \ldots, s_k\}$ is called a *uniform probability space* if the values $p_1 = p(s_1)$, $p_2 = p(s_2), \ldots, p_k = p(s_k)$ are all equal: $p_1 = p_2 = \cdots = p_k$.

8 Theorem

In a uniform probability space S consisting of k elementary events $s_1, \ldots, s_k$, the probability of any elementary event is $1/k$: $p(s_i) = 1/k \, (i = 1, 2, \ldots, k)$.

Proof. By hypothesis, S is a uniform probability space. Hence

$$p_1 = p_2 = \cdots = p_k = p$$

where we have set each $p_i = p$ (the common value). By Equation 1.5 we have

$$p_1 + \cdots + p_k = 1$$

Hence

$$p + \cdots + p = 1$$

$$kp = 1$$

$$p = \frac{1}{k}$$

This completes the proof.

This theorem is the basis for the computation of various probabilities. For example, in the coin-tossing experiment, $S = \{H, T\}$. Thus $k = 2$ here, and $p(H) = \frac{1}{2}$, $p(T) = \frac{1}{2}$. Similarly, if we toss a die, we may take $S = \{1, 2, 3, 4, 5, 6\}$. Hence $k = 6$ and $p = \frac{1}{6}$. Thus the probability of tossing a 4 is $\frac{1}{6} = .167$. If we choose 1 card at random from a shuffled deck, we may choose S to be the set of possible cards. Hence $k = 52$, and the probability of choosing (say) the 10 of hearts is $\frac{1}{52} = .0192$. The same probability applies to each of the possible cards.

Underlying these elementary computations is the basic assumption of

Theorem 8: The probability space is *uniform*. This is surely a reasonable assumption for cards. For example, why should anyone expect the draw of a 7 of hearts to be more (or less) likely than the draw of a 3 of clubs? The obvious assumption to make is that all cards are equally likely to be drawn. Similarly, if a die is thrown, it seems reasonable to assume that a 3 is just as likely to turn up as a 4, etc. On the other hand, the reader must beware of falling into the Aristotelian trap.[5] Nature cannot be forced. Maybe there *is* a bias against drawing an ace of spades! Certainly it seems likely that in the throw of a die, all faces are *not* equally likely. Indeed, the die is not perfectly symmetrical. (The basis of "loading" dice is to make this asymmetry more pronounced.) Similarly, a coin is not perfectly symmetrical.

Despite these cautionary statements, we shall usually assume that for coins, cards, and dice, the appropriate probability space is uniform. (We then say that we are dealing with an *ideal* or *fair* coin, pack of cards, or die.) The justification is as follows. First, as far as the mathematics is concerned, it often simplifies the theory considerably. Second, the physical reality (based on experiments) is fairly close to this assumption. Finally, the knowledge of an *ideal* coin, die, etc., is obviously useful if we wish to study any coin, die, etc., to find out how far from the ideal it is. Thus our reasons are analogous to the ones that the Greek physicist–geometer might have used when he started out considering lines before curves.

EXERCISES

1. State whether each of the following are probability spaces. Give reasons.

a.

s	A	B	C	D	E
$p(s)$	.1	.2	.3	.4	.5

b.

s	1	2	3	4
$p(s)$	.4	.4	.2	.0

c.

s	u	v	w
$p(s)$	$\frac{1}{6}$	$\frac{1}{3}$	$\frac{1}{2}$

---

5 After Aristotle, who "proved" that nature acts in certain ways. Nature refused to comply.

2. Suppose that $S = \{A, B, C, D\}$ is a probability space, with $p(A) = .1$, $p(B) = .4$, $p(C) = .3$. Find $p(D)$.

3. A uniform space consists of the elements #, !, *, }, and 1. Find $p(!)$.

4. A uniform probability space consists of all the letters of the English alphabet. Evaluate $p(K)$.

5. Let $S = \{A, B, C\}$. Suppose that B is twice as likely as A and that B and C are equally likely. Find $p(A)$, $p(B)$, and $p(C)$.

6. Let $S = \{a, b, c, d\}$. Suppose that b is twice as likely as a, c is twice as likely as b, and d is twice as likely as c. Find $p(a)$.

7. Suppose an experiment has possible outcomes $s_1, \ldots, s_k$ and that this experiment is repeated N times. Suppose these outcomes occur with relative frequencies $f_1, \ldots, f_s$, respectively.
  a. Are $f_1, \ldots, f_k$ the statistical probabilities of $s_1, \ldots, s_k$, respectively?
  b. If we *define* $p(s_i) = f_i$, is S a probability space? Explain.

8. Suppose you tell someone to pick a number (i.e., an integer) from 1 to 10 inclusive and to pick it so that you will not be able to guess the number. What number will he choose? It seems clear that we have a sample space consisting of the 10 numbers 1 through 10. Do you think this is a uniform probability space? Explain why. Devise an experiment and try it on 30 people to check your conjecture.

9. The hearts are removed from a pack of cards and a card is chosen at random[6] from the remaining deck. Find $p(3 \text{ of clubs})$.

10. A card is to be drawn from a well-shuffled deck to determine the statistical probability of drawing a black card. One possible sample space is {black, red}. Is it also possible to use {hearts, diamonds, clubs, spades}? Explain. Can {black, red} be used as a sample space if it is desired to determine the statistical probability of drawing a spade? Explain.

3 EVENTS

The results of an experiment can give facts about the relative frequency of occurrences that are not considered elementary events or outcomes. For example, suppose an experiment is performed several times to determine the suit (hearts, diamonds, clubs, or spades) of a card drawn randomly out of a deck. It is natural to take as the sample space $S = \{\text{He, Di, Cl, Sp}\}$. Suppose the results of the experiment are summarized in Table 1.9. (Here we

---

6 The term "at random" means that the obvious probability space is taken to be uniform.

1.9 Suit of a Card

s	He	Di	Cl	Sp
n	56	43	53	48
f	.280	.215	.265	.240

use n to designate frequency and f to denote relative frequency.) Then it would be a very simple matter to use the results of this experiment to find the relative frequency of obtaining a *red* card. In fact, the red cards are the hearts or the diamonds. Hence a red card occurred $n_{He} + n_{Di} = 56 + 43 = 99$ times out of $N = n_{He} + n_{Di} + n_{Cl} + n_{Sp} = 56 + 43 + 53 + 48 = 200$ times. Thus the relative frequency of a red card is $99/200 = .495$. (As we shall soon see, this result could have been obtained directly by adding the relative frequencies f_{He} and f_{Di}.) Thus we may write red $= \{$He, Di$\}$ and

$$f_{red} = f_{He} + f_{Di} \tag{1.6}$$

We now generalize this example to arbitrary experiments.

9 Definition

Suppose a probability experiment has the possible outcomes $s_1, \ldots, s_k$. An *event* A is any subset of these k outcomes: $A = \{t_1, \ldots, t_p\}$, where the t_i's are different outcomes. We say that A *occurs* in an experiment if one of the outcomes in the subset A occurs. If the experiment is repeated N times, the *frequency* of A is the number of times that A occurs. The *relative frequency* of A [written $f(A)$] is defined by the formula

$$f(A) = \frac{n}{N} \tag{1.7}$$

where n is the frequency of A.

10 Theorem

Let $A = \{t_1, \ldots, t_p\}$ be an event. Suppose that the probability experiment is repeated N times and that t_1 occurs with relative frequency $f(t_1)$, etc. Then

$$f(A) = f(t_1) + f(t_2) + \cdots + f(t_p) \tag{1.8}$$

In brief, the relative frequency of A is the sum of the relative frequencies of the outcomes that constitute A.

Proof. Suppose that t_i occurs m_i times $(i = 1, 2, \ldots, p)$. Then, by Definition 2,

$$f(t_i) = \frac{m_i}{N} \qquad (i = 1, 2, \ldots, p)$$

But if n is the frequency of A, clearly n is the sum of the frequencies of the t_i's. Thus

$$n = m_1 + \cdots + m_p$$

Dividing by N we obtain

$$\frac{n}{N} = \frac{m_1}{N} + \cdots + \frac{m_p}{N}$$

or

$$f(A) = f(t_1) + \cdots + f(t_p)$$

This is the result.

Thus in Table 1.9 the relative frequency of red can be computed using Equation 1.6, which is a special case of Equation 1.8.

11 Example

Using Table 1.6, find the relative frequency of the event A: "a 6 turns up on or before the fourth toss."

Here s_1 was the event "a 6 turns up on the first toss," s_2 the event "a 6 turns up first on the second toss," etc. Thus

$$A = \{s_1, s_2, s_3, s_4\}$$

and the above theorem permits the computation

$$\begin{aligned} f(A) &= f(s_1) + f(s_2) + f(s_3) + f(s_4) \\ &= .150 + .117 + .117 + .200 = .584 \end{aligned}$$

using the results of that table. We may expect a small error because the figures are not exact but rounded off to 3 decimal places. The actual relative frequency is

$$f(A) = \frac{9+7+7+12}{60} = \frac{35}{60} = .583$$

to 3 decimal places.

It seems reasonable to compare the relative frequency of A with the *probability* of A. Strictly speaking we have not as yet defined the probability of an event, but in view of Theorem 10 it is reasonable to add the probabilities of the sample points constituting that event. Using the figures of Table 1.6 we obtain

$$p(A) = .167 + .139 + .116 + .096 = .518$$

where the last figure is in doubt because of roundoff error.

We now formally define the *probability* of an event.

12 Definition

Let $S = \{s_1, \ldots, s_k\}$ be a probability space. An *event* A is any subset $\{t_1, \ldots, t_p\}$ of S. The probability of A, written $p(A)$, is defined by the formula

$$p(A) = p(t_1) + \cdots + p(t_p) \tag{1.9}$$

Definitions 9 and 12 illustrate an important mode of procedure: We try to define a mathematical concept in terms of the corresponding physical phenomenon. Thus Equation 1.9 was obviously motivated by Equation 1.8. For if relative frequencies of sample points approximate probabilities of elementary events, as we originally intended, then (comparing Equations 1.8 and 1.9) the relative frequency of an event will approximate its probability.

Equation 1.9 may be written in the somewhat forbidding form

$$p(A) = \sum_{s \in A} p(s) \tag{1.10}$$

[*Read:* $p(A)$ equals the sum of $p(s)$ for s in A.] Here Σ is the symbol universally used to designate a sum. The term $p(s)$ following the Σ sign is the general expression for a term to be added. The statement "$s \in A$" under the Σ sign is a restriction on the terms $p(s)$ to be added. Thus we add only those terms $p(s)$, where s is in the set A.

Definition 12 takes on an interesting and useful form when S is a uniform probability space. In this case, the probability of a sample point is known (Theorem 8), and therefore the probability of an event can be explicitly computed. The result is given below in Theorem 14, after we introduce an important notation.

13 Notation
If A is any finite set, then the number of elements of A is denoted $n(A)$.

14 Theorem
Let S be a uniform probability space, and let A be an event of S. Then

$$p(A) = \frac{n(A)}{n(S)} \tag{1.11}$$

Proof. Suppose $S = \{s_1, s_2, \ldots, s_k\}$, while $A = \{t_1, t_2, \ldots, t_p\}$. By Definition 12,

$$p(A) = p(t_1) + \cdots + p(t_p)$$

By Theorem 8, $p(t_i) = 1/k$ for $i = 1, 2, \ldots, p$. Thus

$$p(A) = \frac{1}{k} + \cdots + \frac{1}{k} \quad (p \text{ summands})$$

$$= \frac{p}{k}$$

But $n(A) = p$ and $n(S) = k$, by definition. Thus

$$p(A) = \frac{n(A)}{n(S)}$$

and the proof is complete.

Historically, Theorem 14 was taken to be the mathematical *definition* of probability. The classical definition was phrased as follows: *If there are x possible outcomes, all equally likely, and if an event can occur in any one of y ways, then the probability of this event occurring is y/x.* It is seen that this is entirely equivalent to Theorem 14.

Theorem 14 is a device with which many probabilities can be computed. All that is necessary to compute the probability of an event is that the probability space be uniform. However, it is first necessary to find $n(A)$ and $n(S)$. This procedure is of course called *counting*. It is by no means an easy matter if S is a fairly large set. For example, suppose we attempt to find the probability that a word, chosen at random from a standard dictionary, has 5 letters. We might interpret this problem as follows: S = the set of all words in the dictionary under consideration, made into a uniform probability space. (This is our interpretation of the word "random.") A = the set of 5-letter words in that dictionary. To compute $p(A)$ we use Equation 1.11 to find

$$p(A) = \frac{n(A)}{n(S)}$$

so it is merely necessary to count the number of 5-letter words, and the number of words, and finally to divide the former number by the latter one—a tedious procedure! On the other hand, some counting is fairly routine. For example, if 1 card is chosen from a deck of cards, we may easily find the probability that it is a picture card (jack, queen, or king). Here S = set of cards, and hence $n(S) = 52$. The event A = set of picture cards. Since there are 4 jacks, 4 queens, and 4 kings, we have $n(A) = 12$. Hence (assuming a uniform space), the required probability is

$$p(A) = \frac{12}{52} = \frac{3}{13} = .230$$

In everyday language, there are 12 chances out of 52, or 3 out of 13, or about 23 out of 100 of choosing a picture card.

In an experimental situation the choice of a sample space is somewhat arbitrary. In the above situation, it would have been valid to choose as the sample space the various ranks: $S = \{\text{ace}, 2, 3, 4, 5, 6, 7, 8, 9, 10, J, Q, K\}$. The event A would be $\{J, Q, K\}$. Since it is "reasonable" to suppose S uniform, we would have $p(A) = \frac{3}{13}$ directly. On the other hand, we might have chosen S to be the simpler sample space: $S = \{\text{picture, no picture}\}$. However, it is "unreasonable" to assume S to be uniform, and we cannot apply the simple formula 1.11.

We also point out that our definition of an event is broad, because it includes every possible subset of S. In particular, it is understood to include the *empty set* $\emptyset$ (a convenient set that includes no elements at all). $\emptyset$ is also called the *impossible event*. In this case we interpret Definition 12 (Equation 1.9) to include the equation

$$p(\emptyset) = 0 \tag{1.12}$$

as a special case. This definition is certainly a reasonable one from the point of view of relative frequency. Indeed, the impossible will not happen, and its relative frequency is 0. Another event of passing interest is the event S itself. S is also called the *certain event*, because S will clearly occur. We have

$$p(S) = 1 \tag{1.13}$$

This is also seen to be true from a relative-frequency point of view. Mathematically, the reader should convince himself that this is a consequence of Definition 12 and Equation 1.5.

Another special case is an event consisting of only one elementary event: $A = \{s_i\}$. Equation 1.9 implies, in this case, that $p(A) = p(s_i)$. Although there is a logical distinction between an element and a set that contains only that element, usage and notation blur that distinction with very little, if any, confusion. It makes little difference whether we regard the drawing of an ace of spades as a sample point or as an event. The probabilities are the same.

Finally, we make some remarks on usage. In the real world, or even the mathematical one, we are seldom presented with the simple problem "Find the probability of an event A." Rather, it is customary and convenient to use some circumlocution. If a set A is given, we often ask: What is the probability that a sample point s is in A? This means: What is the value of $p(A)$? Similarly, in the above illustration, the question "What is the probability that the chosen card is a picture card?" was answered by first defining the set A of picture cards and then finding $p(A)$. In the same way we find the "probability of tossing an even number on 1 die" by first defining the uniform probability space $S = \{1, 2, 3, 4, 5, 6\}$ and the set A of even numbers $(A = \{2, 4, 6\})$ and then finding $p(A) = n(A)/n(S) = \frac{3}{6} = \frac{1}{2}$.

EXERCISES

1. Let $S = \{X, Y, Z\}$. List all the possible events. (Do not neglect the empty set.)

2. Let S be the probability space whose table is as follows:

s	a	b	c	d	e
p	.2	.1	.3	.1	.3

a. Find the probability that s is a vowel. What is the associated event A?

b. What is the probability that s is a, b, or c? What is the associated event A?

3. Referring to Table 1.3 find the probability that when 3 dice are thrown, the highest number showing is even. What is the associated event? Compare your figure with the relative frequency of even numbers in each series of 100 experiments.

4. Referring to Table 1.6 find the probability that 9 or more tosses of a die are required before a 6 turns up. Give the associated event. Compare with the relative frequency in that table.

5. Using the results in the table of Exercise 2, Section 1, find the probability that when 2 dice are thrown, the sum of the numbers turning up is 7 or 11. Give the associated event. Compare with the relative frequency as indicated in the table. Also, find the probability of tossing a sum that is either 6, 7, or 8.

6. A card is drawn at random from a standard deck of cards. What is the probability that the card is an ace and/or a spade? List the sample points in the associated event.

7. A number (i.e., an integer) is chosen at random from 1 to 100 inclusive. What is the probability that the digit 9 appears in that number? List the sample points in the associated event.

8. A license-plate number begins with either a letter or a number. Assuming that the sample space of letters and numbers is uniform and that the number 0 and letter O are distinguishable from each other, what is the probability that a license plate begins with a number?

9. An integer is chosen at random between 1 and 21 inclusive. Find the probability that

a. it is less than 10.   **b.** it is divisible by 3.

c. it is divisible by 5.   **d.** it is divisible by 3 but not by 6.

e. it is divisible by 7 but not by 3.

10. Suppose S is the set of integers from 1 through 100. Let $A =$ the even integers of S, $B =$ the integers divisible by 7 in S, $C =$ the integers less than or equal to 10 in S, and $D =$ the perfect squares in S. Find

a. $n(S), n(A), n(B), n(C), n(D)$.

b. $p(S), p(A), p(B), p(C), p(D)$.

Assume that S is a uniform probability space.

11. Do Exercise 9 if the integer was chosen from 1 to 100.

12. Do Exercise 9 if the integer was chosen from 1 to 1,000.

13. Let $S = \{x, y, x, u, v\}$ be a sample space as follows:

s	x	y	z	u	v
p	.1	.3	.1	.4	.1

    a. Let $A = \{x, y, u\}$. Find $n(A)$. Find $p(A)$.
    b. Compute $\Sigma_{s \in \{u,v\}}\, p(s)$.
    c. Write $p(x) + p(y) + p(v)$ using the "Σ" notation.

14. A certain radio station only plays the "top 25" hit tunes, and they are played at random. However, half of the air time is devoted to random announcer chatter (humor, advertising, station identification, etc.). Set up a sample space for what you will hear when you turn on the radio. What is the probability that you will hear a song in the "top 10"? (Assume, for simplicity, that each tune lasts 3 minutes.)

4 SOME EXAMPLES

We now consider some examples that will illustrate some of the ideas of the preceding sections.

15 Example

When 2 dice are thrown, what is the probability that the sum of the numbers thrown is 5?

    The natural sample space to choose is the possible ways the 2 dice can fall. This is best illustrated in Fig. 1.10. We call one die A and the other B;

1.10   Sample Space for 2 Dice

A \ B	1	2	3	4	5	6
1						
2			(2, 3)			
3		(3, 2)				
4				(4, 4)		
5		(5, 2)				(5, 6)
6						

the outcomes for A (1, 2, 3, 4, 5, or 6) are put in the first column and those for B are put in the first row. The entries in the figure represent all the possible outcomes of the experiment. Thus the entry (x, y) in row x and column y signifies that A turns up x and B turns up y. *The elementary events consist of all (x, y), where $x = 1, 2, \ldots, 6$ and $y = 1, 2, \ldots, 6$.* Figure 1.10 only lists a few typical entries.

If we imagine that die A is readily distinguished from die B (say A is green, B is red), then we can see that the sample points $(2, 3)$ and $(3, 2)$ are distinguishable outcomes. Furthermore, it seems reasonable to suppose that the 36 outcomes are equally likely. Indeed, why (for example) should $(4, 6)$ be any more, or less, likely than $(2, 3)$? This is no proof, of course, because the ultimate proof is experimental.[7] Thus we shall assume that the probability space is *uniform*.

If the dice are similar in appearance, then the experimenter will not be able to distinguish (say) $(1, 3)$ from $(3, 1)$. Nevertheless, the dice are distinct and we may assume the sample space to be the uniform space of 36 elements as in Fig. 1.10.

We can now easily answer the question proposed in Example 15. The event "sum $= 5$" consists of the elementary events $\{(1, 4), (2, 3), (3, 2), (4, 1)\}$. There are 4 such points. Also (see Fig. 1.10), there are $36 = 6 \times 6$ elementary events in S. Using Theorem 14 we have

$$p(\text{sum is } 5) = \tfrac{4}{36} = \tfrac{1}{9} = .111$$

(Compare with Exercise 2 of Section 1, where this probability was stated without proof.)

We can also easily find the probability that the sum on the dice is any of the possibilities $2, 3, \ldots, 12$. Referring to Fig. 1.11, where the elementary

1.11   Two-Dice Sample Space, with Events Corresponding to the Various Sums

7 In Section 5 of Chapter 3, however, we shall consider this assumption in more detail.

events are illustrated as dots, the associated events $A_2, A_3, \ldots, A_{12}$ are seen to contain the number of elementary events as indicated in Table 1.12. This table also gives the probabilities reduced to lowest terms, and the probabilities in percent.

1.12   Sum on 2 Dice

Sum on 2 dice	2	3	4	5	6	7	8	9	10	11	12
Number of sample points	1	2	3	4	5	6	5	4	3	2	1
Probability	$\frac{1}{36}$	$\frac{1}{18}$	$\frac{1}{12}$	$\frac{1}{9}$	$\frac{5}{36}$	$\frac{1}{6}$	$\frac{5}{36}$	$\frac{1}{9}$	$\frac{1}{12}$	$\frac{1}{18}$	$\frac{1}{36}$
Probability (percent)	2.8	5.6	8.3	11.1	13.9	16.7	13.9	11.1	8.3	5.6	2.8

16   Example

Five coins are tossed. What is the probability that exactly 2 heads turn up?

If the coins are called A, B, C, D, and E the natural sample space consists of all the distinguishable occurrences of heads and tails. Letting HHTHH denote the occurrence of H, H, T, H, and H on A, B, C, D, and E, respectively, etc., we can easily construct the sample space of all possible combinations of heads and tails. In alphabetical order (reading downward from the first column) the sample space S is given in Fig. 1.13.

1.13   Sample Space for 5 Tossed Coins

HHHHH	HTHHH	THHHH	TTHHH
HHHHT	HTHHT	THHHT	√ TTHHT
HHHTH	HTHTH	THHTH	√ TTHTH
HHHTT	√ HTHTT	√ THHTT	TTHTT
HHTHH	HTTHH	THTHH	√ TTTHH
HHTHT	√ HTTHT	√ THTHT	TTTHT
HHTTH	√ HTTTH	√ THTTH	TTTTH
√ HHTTT	HTTTT	THTTT	TTTTT

By actual count, S contains 32 elements: $n(S) = 32$. If we let $A_2 =$ the set of elementary events with exactly 2 heads occurring, we find $n(A_2) = 10$ by counting. (The elements of A_2 are checked in Fig. 1.13.) Thus $p(A_2) = n(A_2)/n(S) = \frac{10}{32} = .312$. Here we are assuming that S is a uniform space—i.e., that each of the outcomes in Fig. 1.13 is equally likely. This seems reasonable enough, although a "proof" must be obtained empirically. As with the dice experiment of Example 15, we shall analyze this assumption in greater detail in a later section.

By further inspection of Fig. 1.13, we can find the probabilities of no heads, 1 head, etc. These are listed in Table 1.14.

1.14   Number of Heads Among 5 Tossed Coins

Number of heads	0	1	2	3	4	5
Number of sample points	1	5	10	10	5	1
Probability	$\frac{1}{32}$	$\frac{5}{32}$	$\frac{10}{32}$	$\frac{10}{32}$	$\frac{5}{32}$	$\frac{1}{32}$
Probability (percent)	3.1	15.6	31.3	31.3	15.6	3.1

It is worth noting that if we were to run an experiment to determine the relative frequencies of no heads, 1 head, etc., we should probably choose as our sample space the 6 outcomes $s_0, s_1, \ldots, s_5$. The advantage, however, of using the 32 outcomes of Fig. 1.13 over these 6 outcomes is apparent in our theoretical analysis. We chose the 32 occurrences of Fig. 1.13 because we had reason to believe that this was a *uniform space*, and we were thus able to compute probabilities. When the larger uniform space was used, our "outcome" s_2 was reinterpreted as an "event" A_2.

17   Example

A player has 2 coins and plays the following game. He tosses 1 coin. If a head occurs, he wins. If not, he tosses the other coin, and if a head occurs then, he also wins. But if not, he then loses the game. What is his probability of winning?

A natural sample space would be all the outcomes H, TH, TT. Of these, the first two constitute the event of winning. Hence we might say that

$$p(\text{winning}) = \tfrac{2}{3}$$

But we may also reason that if the player wins on his first try, it does him no harm to toss the second coin just to see what happens. The sample space is then HH, HT, TH, TT. In this case there are *three* ways of winning and *four* possible outcomes. Hence we might say that

$$p(\text{winning}) = \tfrac{3}{4}$$

Which is right? We choose the latter result because we assume that the 4-point space is uniform (all outcomes are equally likely), and hence we may *correctly* use Theorem 14, which permits us to compute probabilities by counting. The first space is *not* uniform. Indeed, most people would bet on H rather than on TT. A good empirical way of checking this result is to play this game many times and find the relative frequency of winning.

Note that we obtained a uniform space by means of an artifice. We thought of H as 2 outcomes HH and HT.

This example is interesting historically, because the famous and respected mathematician D'Alembert claimed that the probability was $\frac{2}{3}$ and stuck to his guns for a long while. If the reader falls into the trap of assuming that any probability space is uniform, he now has the knowledge that it happened before.

18 Example

Of the 10 families living on a certain street, 7 are opposed to air pollution and 3 are either in favor or undecided. A polltaker, not knowing where people stand on this issue, chooses 2 families at random. What is the probability that both families he chose are opposed to air pollution? What is the probability that neither family opposes air pollution?

If we label the families that oppose pollution A, B, C, D, E, F, G and the families that are in favor or undecided H, I, J, then we have the sample space of Fig. 1.15. Here the situation is similar to the 2-dice space (Fig. 1.10) except that the diagonal (A,A), (B,B), etc., is excluded, because the

1 \ 2	A	B	C	D	E	F	G	H	I	J
A	X									
B		X								
C			X							
D				X						
E					X					
F						X				
G							X			
H								X	•	•
I								•	X	•
J								•	•	X

1.15   Sample Space for Choosing 2 Families from 10

polltaker knows better than to bother a family twice. Here, for example, (F, A) is in row F, column A, and represents interviewing F first and then A. There are $10 \times 10 = 100$ squares, of which the main diagonal (upper left to lower right) has been excluded. Hence we have a total of $10^2 - 10 = 90$ sample points. The lightly shaded squares represent the event "OPPOSED" that the 2 families interviewed were opposed to air pollution. We readily see that n (OPPOSED) $= 7^2 - 7 = 7(7-1) = 42$. Therefore, the probability that both families will be opposed to air pollution is

$$p \text{ (OPPOSED)} = \tfrac{42}{90} (= 46.7\%)$$

The probability that both families will be in favor or undecided is, using the squares with dots,

$$p \text{ (POLLUTE)} = \tfrac{6}{90} (= 6.7\%)$$

19 Example

From our point of view, an urn is a bowl, box, or receptacle containing marbles, coins, pieces of paper, or other objects that feel the same to a blindfolded person but have certain identifiable characteristics. Using an urn, we obtain a very realistic physical model of a uniform probability space. For example, if 15 billiard balls (numbered 1 through 15) are put in a box and shaken, then when we choose one of these balls without looking, we are reasonably certain that each ball is equally likely to be chosen. The same principle applies to playing bingo or to a drawing in a lottery. A good way for the polltaker of Example 18 to decide which 2 families to choose is to put the names of the 10 families on pieces of paper, mix thoroughly, and choose 2 slips of paper.

Using urns, we can even create nonuniform probability spaces. For example, if an urn contains 7 green marbles, 5 red marbles, and 1 blue marble, we may label the marbles conveniently as follows:

$$g_1, g_2, g_3, g_4, g_5, g_6, g_7, r_1, r_2, r_3, r_4, r_5, b_1$$

Then the events G (green), R (red), and B (blue) are defined in the natural manner: $G = \{g_1, g_2, \ldots, g_7\}, R = \{r_1, r_2, \ldots, r_5\}, B = \{b_1\}$. The probabilities for these events are governed by Theorem 14. Hence $p(G) = \tfrac{7}{13}, p(R) = \tfrac{5}{13}, p(B) = \tfrac{1}{13}$, and we may regard this model as a 3-point, nonuniform model. In much the same way, any finite probability space may be realized by an urn model, as long as the probabilities are rational numbers (i.e., fractions). Since irrationals may be closely approximated by rationals, we can certainly approximate any finite sample space by an urn model. This is a useful idea, for it shows that it is not too unrealistic to consider uniform probability spaces — all finite sample spaces may be regarded as approximations of these.

EXERCISES

1. Using the (uniform) probability space of Fig. 1.10, find the probability that, when 2 dice are thrown, at least one 4 shows up. Indicate the associated event on a diagram.

2. As in Exercise 1, find the probability that when 2 dice are thrown, at least one 1 or 6 shows up. Draw a diagram of this event.

3. As in Exercise 1, find the probability that the difference between the high and low number thrown is 2 or more.

4. In analogy with the 3-dice experiment (Example 1 of Section 1), find the probability that, when 2 dice are tossed, the highest number thrown is 6. Similarly, find the probability that the high number is, 5, 4, 3, 2, and 1. Sketch the associated events in a single diagram.

5. Using the (uniform) probability space of Fig. 1.13, find the probability that, when 5 coins are tossed consecutively,
  a. the first coin lands heads.
  b. the first 2 coins land heads.

6. As in Exercise 5, find the probability that when 5 coins are tossed consecutively, a run of 3 or more heads in a row occurs. List the sample points of the associated event.

7. As in Exercise 5, find the probability that when 5 coins are tossed consecutively, the sequence HTH occurs somewhere during the run.

8. As in Exercise 7, find the probability that either HTH or THT occurs somewhere during the run.

9. Three coins are tossed. List the sample points. Find the probability of
  a. no heads. **b.** 1 head. **c.** 2 heads. **d.** 3 heads.

10. Was D'Alembert originally correct? Play the game of Example 17 *one hundred times* and count the number of wins (W) and losses (L). (Do *not* use the artifice of the unneeded toss.) Compute the relative frequency, and compare with .667 (incorrect) and .750 (correct). Suppose you decide that a relative frequency of .75 or more verifies the text answer of .750, and a relative frequency of .67 or less verifies the (allegedly incorrect) answer. Which answer, if any, have you verified?[8]

11. Alex, Bill, Carl, Dave, Emil, and Fred are members of a club. They

8 We have probabilities within probabilities here. Assuming that .75 is the correct probability, we shall later learn how to find the probability that in a run of 100 experiments, 75 or more wins occur, and similarly that 67 or less occurs. The same probabilities can be computed on the

decide to choose the president and vice president at random. (They use an urn, of course.) Using the technique of Example 18, find the probability that
   a. Carl is an officer.
   b. Alex and Bill are not officers.
   c. either Dave or Emil is president.
   d. Fred is president and/or Bill is vice president.
   e. Dave resigns from the club. (He will, if Bill becomes president.)

   12. An urn contains 6 white balls and 1 black ball. Two balls are drawn at random. What is the probability that both are white? What is the probability that the black ball is chosen?

   13. A gambling house offers the following game for the amusement of its customers. Two dice are tossed. If two 6's turn up the customer wins $10.00. If only one 6 turns up, the customer wins $1.00. If no 6's turn up, he loses $1.00. Find the respective probabilities of winning $10.00, winning $1.00, and losing $1.00.

   14. Gary and Hank are evenly matched rummy players. They engage in a championship match in which the best 3 out of 5 wins the championship. Once either player has won 3 games, the tournament ends and that player is declared the winner. Set up a probability space. Be sure you state the probabilities for each elementary event.

5 SOME GENERALIZATIONS

We have thus far set the stage for a theoretical study of probability using the notion of a *probability space* as the unifying idea behind random events. Although the motivating force for the definition was the idea of statistical

---

assumption that the probability of a win is $\frac{2}{3}$. This is summarized in the following table:

	Probability of 75 or more wins	Probability of 67 or less wins	Probability of no determination
If $p = \frac{3}{4}$	55%	4%	41%
If $p = \frac{2}{3}$	4%	57%	39%

   Thus, even assuming that $p = \frac{3}{4}$, about 4 percent of all people doing this problem will find 67 or less wins and will decide $p = \frac{2}{3}$. Note that 41 percent will come to no conclusion. If 1,000 experiments were run, the figures in this table could be sharpened. The above method of decision is not the one that a statistician would choose. But regardless of the statistician's method, he will always have a clause to the effect that with a small probability, the wrong decision will be made.

probability, Definition 6 is so general that it can be used to describe a variety of situations. Let us now consider some of these.

Statistical Results as Sample Space. If an experiment has outcomes $s_1, s_2, \ldots, s_k$ which occur with frequencies $n_1, n_2, \ldots, n_k$, respectively, Definition 2 gives the definition of the relative frequencies $f_1, f_2, \ldots, f_k$ of these outcomes. These relative frequencies satisfy Equations 1.1 and 1.2, which are precisely what probabilities are required to satisfy (Equations 1.4 and 1.5). Thus the outcomes $s_1, \ldots, s_k$ with relative frequencies $f_1, \ldots, f_k$ form a probability space, which we may call the *statistical probability space*, determined by the results of the experiment.

It is not even necessary for the outcomes to be the result of a series of random experiments. Any table giving the number of occurrences of various events can be interpreted in this manner. For example, consider the following (imaginary) information about the students in a certain college:

Sleeping Habits of Students

Sleep over 10 hours per day	30%
Sleep over 8 hours, less than 10	42%
Sleep 8 or less hours per day	23%
Never sleep	5%

Clearly, this may be regarded as a 4-element probability space. The numbers .30, .42, .23, and .05 are called relative frequencies rather than probabilities in order not to mislead the unwary.

Statistical Probability as Sample Space. This is the application originally intended for a sample space to describe. If an experiment has possible outcomes $s_1, \ldots, s_k$, we take as a physical fact that there are numbers $p_1, \ldots, p_k$ satisfying Equations 1.4 and 1.5 which are approximately the relative frequencies of $s_1, \ldots, s_k$ if the experiment is repeated a large number of times. We may say that the probability space postulated in this way is the *limit* of statistical sample spaces as in the case above.

Finite Sets as Sample Space. If $S = \{s_1, \ldots, s_k\}$ is *any finite set* with k elements, we may make S into a *uniform probability space* by defining $p(s_i) = 1/k$ for $i = 1, \ldots, k$. Clearly Equations 1.4 and 1.5 hold. In this case Theorem 14 holds. Recalling that if A is any set, $n(A)$ is the number of elements in A and that $n(S) = k$, we see by Theorem 14 that

$$p(A) = \frac{n(A)}{k} \tag{1.14}$$

or
$$n(A) = kp(A) \tag{1.15}$$

By making any finite set into a uniform probability space, we may use Equations 1.14 and 1.15 to translate statements about numbers of elements into statements about probabilities, and conversely. For example, the statement "There are 5 even numbers among the first 10 positive integers," concerns $S = \{1, 2, \ldots, 10\}$, $A = \{2, 4, 6, 8, 10\}$, $k = 10$, $n(A) = 5$, and therefore $p(A) = \frac{5}{10} = \frac{1}{2}$. We say interpret this statement as, "The probability of choosing an even number among the first 10 positive integers is $\frac{1}{2}$." It is not necessary to regard this latter statement from the point of view of running a large number of experiments with urns. Rather, by Equation 1.15, it can be interpreted solely in terms of the relative number of even numbers among the numbers of S. However, the probability statement gives us information about, and perhaps a feeling for, the scarcity, or density, of even numbers.

We can give a simple example of each of these three kinds of applications by considering the game of Example 17. Suppose this game is attempted 20 times and is won 16 times (relative frequency $\frac{16}{20} = .80$) and lost 4 times (relative frequency .20). We can apply the 3 notions above to the same set of outcomes as follows:

	W	L
A. Statistical relative frequencies	.80	.20
B. Statistical probabilities	.75	.25
C. Uniform probabilities	.50	.50

The 3 sets of figures are interpreted as follows: Row A gives the relative frequencies of what actually happened in the particular run of 20 experiments. Row B gives the theoretical probabilities. In a very large run of experiments, we may expect the relative frequencies to be near these figures. Row C merely implies that there are 2 outcomes, each outcome representing $\frac{1}{2}$ of the total number of outcomes. In actual practice it will always be clear which application is being used. In Section 4 we used C (finite sets as probability space) and thought of the results as applying to B (statistical probability as probability space).

An important generalization of our notion of a sample space is that of an *infinite discrete* sample space. Definition 6 postulated finitely many (k) outcomes. But in many natural situations infinitely many outcomes are conceivable. This was seen in Example 5, in which a die was tossed until a 6 turned up. The natural outcomes were $s_1, s_2, s_3, \ldots$ indefinitely, because we could not definitely exclude any possibility. We should also include s_∞, the case where no 6 ever turns up. In this example we could be relatively sure that, for example, a 6 would turn up before 1,000 throws. Yet an infinite sample space seemed to be called for. If we do use this infinite space, we can easily interpret the event "6 takes 13 or more tosses before arriving" as the event $A = \{s_{13}, s_{14}, \ldots, s_\infty\}$. Here A is an infinite set. Even so, its probability is fairly small.

This is an example of an *infinite discrete* sample space. Its sample points, while infinite in number, can be enumerated $s_1, s_2, \ldots$.[9] There is one sample point for each positive integer. The general definition is a rather straight-forward generalization of Definition 6.

20 Definition

An *infinite discrete sample space* S is a set of points $s_1, s_2, \ldots$ (one for each positive integer) called sample points. Corresponding to each sample point s_i is a number p_i called the probability of s_i. The numbers p_i must satisfy the conditions

$$0 \leqslant p_i \leqslant 1 \qquad (i = 1, 2, 3, \ldots) \tag{1.16}$$

$$p_1 + p_2 + \cdots = \sum_{i=1}^{\infty} p_i = 1 \tag{1.17}$$

Here Equation 1.17 is an infinite series. The meaning of this equation is that if enough terms are taken in this sum, the finite sum will be as close to 1 as we want it to be. Thus we can make $\sum_{i=1}^{n} p_i$ greater than .999 (and less than or equal to 1.000) by choosing n large enough. Similarly, we can exceed .9999, etc. Therefore, from a probability point of view, the entire sample space is practically concentrated in finitely many sample points.

We may also look upon Equation 1.17 in the following way. If probabilities are desired to 4 decimal places, we need only choose finitely many sample points, say $s_1, s_2, \ldots, s_{100}$, and the probability of the event $\{s_1, \ldots, s_{100}\}$, which is $p_1 + \cdots + p_{100}$, will equal 1 to 4 decimal places. The probability of the event $\{s_{101}, s_{102}, \ldots\}$ will be smaller than .00005 and will appear in our table of probabilities as .0000. If more decimal places are called for, we will need more elementary events, but the principle is the same.

In a later section we shall learn how to compute the probability p_n that n tosses for a coin are required before a head occurs. The formula is $p_n = 1/2^n$. Thus $p(\mathrm{H}) = \frac{1}{2}$, $p(\mathrm{TH}) = \frac{1}{4}$, $p(\mathrm{TTH}) = \frac{1}{8}$, etc. A table (to 2 decimal places) is as follows:

Number of tosses	1	2	3	4	5	6	7	8 or more
Probability	.50	.25	.13	.06	.03	.02	.01	.00

9 In the dice experiment, the addition of one extra sample point s_∞ did not "increase the number of sample points." For example, we may renumber $s_\infty, s_1, s_2, \ldots$ as $t_1, t_2, t_3, \ldots$. ($t_1 = s_\infty$, $t_2 = s_1$, etc.)

Thus an infinite sample space is reduced to one having 8 sample points. More decimal places will require more sample points (and give better accuracy).

Sample spaces even more complicated than discrete ones are possible, although we do not define them in this text.

EXERCISES

1. New York City is divided into 5 boroughs. The population (1960 census) was as follows:

Population of New York City (1960 Census), by Boroughs

Borough	Population (in thousands)
Manhattan	1,698
Bronx	1,425
Brooklyn	2,627
Queens	1,810
Richmond	222

Use this table to make a probability space of the boroughs.

2. The following is a summary of a school budget for a city:

A School Budget

Account	Percentage of total
Salaries	74.44
Debt service	5.35
Instructional supplies	5.44
Maintenance	2.26
Capital outlay	2.24
Transportation	1.91
Operational expense	1.82
Contingency fund	1.78
Other	4.76
Total	100.00

Is this a probability space? Does this mean that the money is spent at random? Explain.

3. The probability that an integer between 1 and 100 inclusive is a prime number is .25. Restate this fact without using the language of probability.

4. It is desired to choose a positive integer at random so that each positive integer is as likely to be chosen as another. Show that this cannot be done. In brief, show that there is no such thing as a uniform, infinite, discrete sample space.

CHAPTER 2 COUNTING

INTRODUCTION

Many important problems in probability theory involve sets with a rather large number of elements. Therefore, to compute probabilities it is necessary to learn how to count efficiently. The reader already knows how to do this in many cases. If a room is 12 by 13 ft, and if a square tile is 1 by 1 ft, then it is not necessary to count, one by one, how many squares are needed to tile the room. The answer, of course, is $12 \times 13 = 156$ tiles. In this chapter we shall learn several short cuts of this type and apply them to probability problems.

The proper framework of counting theory is within the theory of sets. Indeed, counting is a way of assigning to any set A the number $n(A)$ of elements in this set. We count things. Hence sets — collection of things — are needed to understand counting.

1 PRODUCT SETS

Product sets arise when the objects in a given set can be determined by two characteristics. A familiar example is a pack of cards. A card is determined by the suit (hearts, diamonds, clubs, or spades) and by the rank (A, 2, 3, 4, 5, 6, 7, 8, 9, 10, J, Q, K). Our way of describing a card (e.g., a 5 of hearts) makes this explicit. The set of cards can be pictured as in Fig. 2.1. Thus the set

2.1 Cards

	A	2	3	4	5	6	7	8	9	10	J	Q	K
Hearts													
Diamonds													
Clubs													
Spades													

of cards is determined by two much simpler sets – the set of suits and the set of ranks.

We have seen a very similar situation in Example 1.15 (Fig. 1.10). Here the possible outcomes of a tossed pair of dice was also determined by 2 characteristics – the number on the first die and the number on the second. Thus the possible outcomes were determined by 2 much simpler sets – the set $A = \{1, 2, 3, 4, 5, 6\}$ and $B = \{1, 2, 3, 4, 5, 6\}$. (In this case the sets were identical.) This situation is greatly generalized in the following definition.

1 Definition

If A and B are 2 sets, the *product set* $C = A \times B$ is defined as the set of all ordered couples (a, b), where a is in A and b is in B.

Remark. An ordered couple (a, b) consists of 2 components. Its first component is a, its second is b. By definition, 2 ordered couples are equal if they agree in both of their components:

$$(a, b) = (c, d) \qquad \text{if and only if } a = c \text{ and } b = d \qquad (2.1)$$

For example, $(1, 3) \neq (3, 1)$ despite the fact that both of these ordered couples have 1 and 3 as components. It is for this reason that they are called *ordered* couples. The order in which the components are listed is a significant factor in determining the ordered couple.

When we draw a diagram of the product set $A \times B$, we form a table by putting the elements of A down in a column and B in a row. Then $A \times B$ will be the body of the table, with (a, b) appearing in the row opposite a and column under b. (See Figs. 1.10 and 2.1 for examples.)

Using Definition 1, we can give several additional examples of product sets. If a coin is tossed, the outcomes may be taken as the 2 elements of the set $A = \{H, T\}$. If a coin is tossed twice, the outcomes may be regarded as the elements of $A \times A$ (see Fig. 2.2). If a coin is tossed and a card is chosen from a standard deck, the outcomes may be regarded as the product of

2.2 *Sample Space for 2 Coins*

	H	T
H	HH	HT
T	TH	TT

{H, T} and of C, the set of cards. If we wish to classify automobiles according to color and make, we might choose $A =$ {black, blue, red, green, beige, white, other} and $B =$ {Chevrolet, Ford, Plymouth, Pontiac, Lincoln, Volkswagen, other}. Then $A \times B$ will be an appropriate set to classify cars. (We would expand A and B if we wanted more detail.) In the polltaker example (Example 18 of Chapter 1), we might take $A =$ the set of all 10 families. But (see Fig. 1.15) the set of possible outcomes was *not* $A \times A$, because the diagonal squares [those of the form (a, a)] did not determine an outcome.

The following theorem on counting the elements in a product set has probably been known to the reader since his early childhood.

2 Theorem

For any sets A and B,

$$n(A \times B) = n(A) \cdot n(B) \qquad (2.2)$$

We multiply the number of elements in A and the number in B to obtain the number in $A \times B$. Any proof would be more confusing than looking at Fig. 2.1, where $4 \times 13 = 52$. This theorem is so basic that it is often taken as the definition of multiplication in a theoretical treatment of multiplication.

The situation is entirely similar if the objects in a certain set can be determined by more than two characteristics. In this case we form the product of two or more sets.

3 Definition

If $A_1, \ldots, A_s$ are sets, the *product* $C = A_1 \times A_2 \times \cdots \times A_s$ is defined as the set of ordered s-tuples $(a_1, a_2, \ldots, a_s)$, where a_i is in A_i for $i = 1, 2, \ldots, s$.

Remark. As before, a_1 is called the first component, ..., a_s the sth component of the ordered s-tuple. Two s-tuples are equal if they are equal component by component. A 3-tuple is called a triple and a 2-tuple a couple.

Some examples of s-tuples have been encountered before. In Example 1.16 (5 tossed coins), each outcome was a 5-tuple, where each component was from the set $A =$ {H, T}. The set of outcomes was therefore $C = A \times A \times A \times A \times A$. In analogy with Theorem 2 (see Theorem 4),

$n(C) = 2 \times 2 \times 2 \times 2 \times 2 = 2^5 = 32$, as explicitly noted in Fig. 1.13. If 4 dice are tossed, the possible outcomes may be regarded as elements of $B \times B \times B \times B$, where $B = \{1, 2, 3, 4, 5, 6\}$. Each component is merely the number appearing on the corresponding die. Here *ordered* couples are wanted, because, for example, $(1, 2, 4, 1)$ is to be distinguished from $(4, 1, 2, 1)$.

We may regard a product of 3 sets as the product of the first 2 sets times the third:

$$A \times B \times C = (A \times B) \times C$$

For example, to find all the occurrences when 3 coins are tossed, we choose $A = \{H, T\}$. We then consider the outcomes as completely determined by the outcome on the first 2 coins and on the third coin; i.e., we consider $(A \times A) \times A$. We may systematically proceed as in Fig. 2.3.

2.3 *Three-Coin Sample Space*

	H	T
HH	HHH	HHT
HT	HTH	HTT
TH	THH	THT
TT	TTH	TTT

In analogy with Theorem 2, we have the following useful theorem.

4 Theorem

For any sets $A_1, \ldots, A_s$, we have

$$n(A_1 \times A_2 \times \cdots \times A_s) = n(A_1) \cdot n(A_2) \cdot \ldots \cdot n(A_s) \tag{2.3}$$

The proof is by successive application of Theorem 2, because a product of s sets may be defined in terms of a product taken 2 at a time.

Theorem 4 greatly expands our ability to count; hence the range of probability problems (using uniform spaces) is enlarged. The examples below illustrate some of the possibilities.

In all the examples to follow, the motivation for using product sets is as follows. An outcome is determined by what happens at various "stages." At the first stage, any possibility from set A_1 may occur. Regardless of what happens at the first stage, at the second stage any possibility from set A_2 may occur; similarly for later stages. The outcomes constitute the product set

$A_1 \times \cdots \times A_8$, and the first component of an outcome is what occurred at the first stage, etc.

5 Example

Three dice are tossed. What is the probability that no 6 turns up? That neither a 5 nor a 6 turns up?

Let $A_6 = \{1, 2, 3, 4, 5, 6\}$. The sample space S consists of three numbers (x, y, z), each chosen from A_6. Thus $S = A_6 \times A_6 \times A_6$. Hence $n(S) = 6^3 = 216$. Now let A_5 be the set $\{1, 2, 3, 4, 5\}$ and let E_5 be the event "no 6 turns up"; then E_5 consists of all the triples (x, y, z), where each of x, y, z is in A_5. Thus $E_5 = A_5 \times A_5 \times A_5$. Using Theorem 4, $n(E_5) = 5^3 = 125$. Thus the probability of E_5 may be found using the basic Theorem 14 of Chapter 1:

$$p(E_5) = \frac{n(E_5)}{n(S)} = \frac{5^3}{6^3} = \frac{125}{216} (= .579)$$

Similarly, if A_4 is the set $\{1, 2, 3, 4\}$, the event E_4, "neither a 5 nor a 6 turns up," is the set $A_4 \times A_4 \times A_4$ and has probability

$$p(E_4) = \frac{4^3}{6^3} = \frac{64}{216} (= .296)$$

6 Example

Three dice are tossed. What is the probability that the high number is 6? Similarly, find the probability that the high number is $5, 4, 3, 2,$ or 1.

The method of Example 5 shows that there are $6^3 = 216$ possible outcomes, of which $5^3 = 125$ outcomes had no 6. There are $216 - 125 = 91$ outcomes with a 6, and thus having a high of 6. Similarly, there are $5^3 = 125$

2.4 High Die When 3 Dice Are Tossed

Outcome	Number of outcomes	Probability	Probability (percent)
6 high	$6^3 - 5^3 = 91$	$\frac{91}{216}$	42.1
5 high	$5^3 - 4^3 = 61$	$\frac{61}{216}$	28.2
4 high	$4^3 - 3^3 = 37$	$\frac{37}{216}$	17.1
3 high	$3^3 - 2^3 = 19$	$\frac{19}{216}$	8.8
2 high	$2^3 - 1^3 = 7$	$\frac{7}{216}$	3.2
1 high	$1^3 = 1$	$\frac{1}{216}$	0.5

outcomes with a high of 5 or less (no 6), but $4^3 = 64$ of these outcomes have a high of 4 or less (no 5 or 6). Thus $125 - 64 = 61$ outcomes have a high of 5. Continuing in this way we obtain Table 2.4.

These results were given without proof in Table 1.3.

7 Example

In Fig. 2.5 how many paths can be drawn tracing the alphabet from A through H?

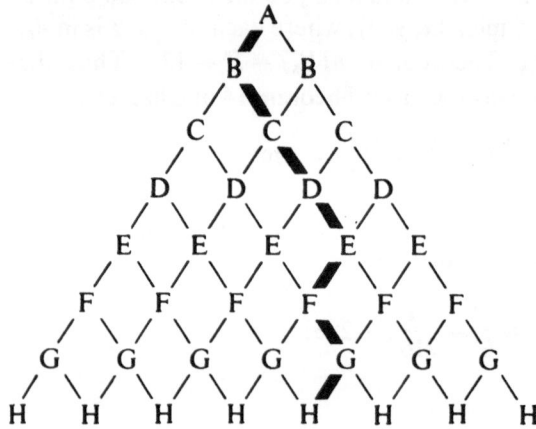

2.5 Alphabet, with Path LRRRLRL

At each state we have a choice of going left (L) or right (R). Furthermore, any sequence of seven directions (L or R) gives a unique path with two different sequences giving different paths. Hence if we let $A = \{L, R\}$, the paths may be regarded as the product set $P = A \times A \times A \times A \times A \times A \times A$, and $n(P) = 2^7 = 128$.

If 7 coins are tossed, the same analysis is involved. To see the correspondence, we may easily imagine starting at A and thinking "heads, we go left; tails, we go right."

8 Example

A roulette wheel has 38 (equally likely) slots, of which 18 are red, 18 black, and 2 green. What is the probability that 5 reds in a row turn up?

The sample space S consists of a sequence (x, y, z, u, v), where each letter represents one of the 38 slots. Thus $n(S) = 38^5$. The event E, "all slots red," consists of 5-tuples in which each entry is 1 of 18 possibilities. Thus $n(E) = 18^5$. Finally, it is reasonable to assume that S is a uniform probability space. Thus

$$p(E) = \frac{n(E)}{n(S)} = \frac{18^5}{38^5} = \left(\frac{18}{38}\right)^5 = \left(\frac{9}{19}\right)^5 (= .0238)$$

9   Example

A pack of cards is divided into 2 piles. One consists only of the 13 spades. The other contains all the other cards. A card is chosen from each pile. What is the probability that a spade picture (J, Q, or K), and a heart is chosen?

If A is the set of spades and B the other cards, we have $n(A) = 13$ and $n(B) = 39$. The sample space is $S = A \times B$, because any element from A may be paired with any element of B. There are $n(A \times B) = 13 \times 39$ elements in S. If we let P be the set of picture spade cards and H the set of hearts, we wish to find the probability of $P \times H$. But $n(P \times H) = n(P) \times n(H) = 3 \times 13$. We naturally assume a uniform space, so we obtain

$$p(P \times H) = \frac{3 \times 13}{13 \times 39} = \frac{1}{13}$$

(Note that it would be foolish to multiply 13 and 39 because an eventual cancellation occurred.)

10   Example

A set has 10 elements in it. How many subsets or events are there?

Call the elements $1, 2, \ldots, 10$. Then a subset is determined by deciding whether 1 is in it or out of it, and similarly for $2, 3, \ldots, 10$. If we let $A = \{\text{in, out}\}$ or $\{\text{I, O}\}$, we see that

$$P = \underbrace{A \times \cdots \times A}_{10\,\text{factors}}$$

may be regarded as the subsets of the given set. For example, I O O O I I I O I O corresponds to the subset $\{1, 5, 6, 7, 9\}$. Thus there are $n(P) = 2^{10} = 1{,}024$ subsets. This includes the whole set, described by I I $\cdots$ I, and the empty set, described by OO $\cdots$ O.

This example clearly generalizes to a set with n elements. Thus, *if S has n element, there are 2^n subsets of S.*

EXERCISES

1. Let $A = \{H, T\}$. In analogy with Fig. 2.3, list all the elements of $A \times A \times A \times A$ in a 4×4 table, by treating this set as the product of $A \times A$ with itself. Using this table, find the probability that, when 4 coins are tossed, 2 heads and 2 tails occur.

2. A menu has a choice of 5 appetizers, 8 main dishes, 4 desserts, and 3 beverages. How many different meals can be ordered if it is required to order one item in each category? How many meals can be ordered if it is required to order a main dish and a beverage but an appetizer and the dessert are optional?

3. Urn A contains 6 green and 5 white marbles. Urn B contains 7 green and 24 white marbles. One marble is chosen at random from each of the urns. What is the probability that both are green? That both have the same color?

4. A die is tossed 4 times in a row. What is the probability that a 6 occurs on the fourth toss but not on the first 3 tosses?

5. Let us define a finite sequence of letters to be a word. How many 3-letter words are there? How many 3-letter words are there which begin and end with a consonant and have a vowel in the middle? (Treat Y as a consonant.)

6. A man decides to exercise by walking 5 blocks daily from a street intersection near his home. Since he is in the middle of a city, once he is at the end of a block, he always has 4 directions to continue his walk. How many paths are possible? Express the set of paths as a product of appropriate sets.

7. Suppose the man in Exercise 6 decides that once he walks a block he will not immediately go back along that same block. He is willing to take his daily 5-block walk as long as he can proceed along a different route every day. Will he stroll for a full year? Express the set of paths as a product of appropriate sets. In particular express the 2 paths in Fig. 2.6 as ordered s-tuples, using your answer.

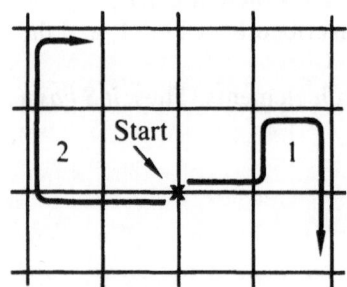

2.6   Two Paths

8. Ten differently colored marbles are to be tossed into 3 urns, A, B, and C. In how many ways can this be done? (It is not necessary to evaluate the answer.) If the marbles are tossed in at random, what is the probability that A will be missed?

9. A multiple-choice test has 3 questions on it, each with 3 possible answers. A class of 30 takes the test. Each student answers all the questions. Show that the teacher will necessarily be able to find 2 identical papers regardless of how uninformed the class is.

10. A house has 5 rooms, and the owner is going to paint the house. He has a choice of 6 colors that he may use in any room. How many color schemes does he have to consider?

11. How many 4-digit numbers are there which do not contain any of the digits 0, 1, or 2? Of these numbers, how many end with the digit 7?

12. A man has a penny, a nickel, a dime, a quarter, and a half-dollar. Using only these coins, how many different amounts can he form?

13. A man has a 1-, a 2-, a 3-, a 5-, and a 10-cent stamp. Explain why product sets *cannot* be used to compute the different amounts of postage possible with these stamps to obtain the same answer as in Exercise 12.

2 MULTIPLICATION PRINCIPLE

In Example 18 of Chapter 1 (the poll of 10 families), product spaces were not immediately usable, because the possibility of polling the same family twice was eliminated. Yet the method of choosing a family was a 2-stage affair. First, one family was chosen (10 possibilities). Then a different family was chosen (9 possibilities). The product $10 \times 9 = 90$ gave the number of possible choices. Here we may say that a choice is an ordered couple (x, y) except that the possibilities for y are determined by the value of x. We now generalize this procedure.

11   Theorem. The Multiplication Principle

Let A and B be sets, and let C be a subset of $A \times B$. Suppose that A has n elements in it, and that for every element a of A there are exactly m elements of the form (a, x) in C. Then

$$n(C) = n \cdot m \qquad (2.4)$$

This theorem is illustrated in Fig. 2.7. Here the set C is indicated by the dots. A has $n = 5$ elements in it, and although B has 10 elements in it, there are only 7 elements of C in each row. Thus $m = 7$, and C has $5 \times 7 = 35$ elements.

2.7

A \ B	b_1	b_2	b_3	b_4	b_5	b_6	b_7	b_8	b_9	b_{10}
a_1			·	·	·	·	·	·	·	
a_2				·	·	·	·	·	·	·
a_3	·			·	·	·	·	·	·	·
a_4	·	·			·	·	·	·	·	·
a_5	·	·	·			·	·	·	·	·

Theorem 11 may be restated in the following useful alternative way.

11′ Theorem

Suppose that an occurrence is determined in 2 stages. If there are n ways in which the first stage can occur, and if for each choice of the first stage, the second stage can occur in m ways, there are $n \cdot m$ possible occurrences.

The proof of Theorem 11 is straightforward. If we call $a_1, a_2, \ldots, a_n$ the elements of A, we have precisely m elements of C that start with a_1; there are m elements of C that start with a_2; etc. In all we have

$$\underbrace{m + m + \cdots m}_{n \text{ times}} = n \cdot m$$

elements of C, which is the result.

As with product spaces, it is essential in this counting process that different ordered couples correspond to different events. For example, if we wish to interview 2 families from among 20, there are $20 \times 19 = 380$ ways this can be done. (At the first stage, we have 20 possibilities. For each of these possibilities, there are 19 ways of selecting the second family.) However, this counting process distinguishes AB from BA. It makes sense to consider which family was interviewed first.

The generalization of Thereom 11 to more than 2 sets is fairly straightforward. For later applications, however, we prefer to generalize Theorem 11′.

12 Theorem

Suppose that an occurrence is determined in s stages. If there are n_1 ways in which the first stage can occur, if for each choice in the first stage the second stage can occur in n_2 ways, and if for each choice in the first two stages the third stage can occur in n_3 ways, etc., then there are $n_1 \cdot n_2 \cdot \ldots \cdot n_s$ possible occurrences.

The proof merely uses Theorem 11 over and over, and we omit the details.

13 Example

How many 3-digit numbers are there in which no digit is repeated?

We think of choosing the number in 3 stages — the first, second, and then the third digit. There are 9 first choices, because we do not allow a number to begin with 0. After the first digit is chosen, there are 9 possibilities for the second, because any of the remaining digits are available. At the third stage, after the first 2 digits are chosen, there are 8 possibilities. Thus there are $9 \times 9 \times 8 = 648$ such numbers.

Problems such as these are sometimes done diagramatically. In Fig. 2.8a, the three digits are blanks to be filled in. The order of filling them is also indicated. We then put the *number of possibilities* in the blanks and multiply as in Fig. 2.8b.

2.8a   **2.8b**

Stage: 1   2   3   Stage: $\boxed{9} \times \boxed{9} \times \boxed{8}$ = 648
                                       1   2   3   Answer

If we were to start this problem with the third digit (the unit's place), we would obtain 10 possibilities for the first stage and 9 for the second. However, the number of possibilities for the third stage (the hundred's place) is sometimes 8 (if 0 has already been chosen) and sometimes 7 (if 0 has not been chosen). Thus Theorem 12 does not apply with this procedure, because the theorem calls for a fixed number of possibilities at the third stage, *regardless* of which elements were chosen at the second stage. Therefore, in any problem involving choices at different stages, we usually take, as the first stage, the event with the most conditions placed on it.

14 Example

How many even 3-digit numbers are there in which no digit is repeated?

In this problem we have conditions on the first digit (not 0) and the last (must be 0, 2, 4, 6, or 8). Whether we start with the first or third digit, we come to an ambiguous situation at the last stage. When this happens (we may call this an "it depends" situation), we proceed as follows. We start at the third digit (the unit's place). There are two possibilities, which we consider separately: 1. The third digit is 0. 2. The third digit is 2, 4, 6, or 8. After disposing of the third digit, we go to the first digit and then to the second. The entire procedure is summarized as follows:

1. Third digit 0. $\quad\quad \boxed{9} \times \boxed{8} \times \boxed{1} = 72$
Stage: $\quad\quad 2 \quad\;\; 3 \quad\;\; 1$

2. Third digit 2, 4, 6, or 8. $\boxed{8} \times \boxed{8} \times \boxed{4} = \underline{256}$
Stage: $\quad\quad\quad 2 \quad\;\; 3 \quad\;\; 1$

Total possibilities $\quad\quad = 328$

In brief, when an "it depends" situation arises, we ask "On what?". We then break up the possibilities accordingly.

An alternative solution proceeds as follows. The *odd* integers with no repetition can be computed according to the scheme

$$\boxed{8} \times \boxed{8} \times \boxed{5} = 320$$
Stage: $\quad 2 \quad\;\; 3 \quad\;\; 1$

By Example 13 there were 648 integers under consideration, so the remaining $648 - 320 = 328$ are even.

15 Example

Three cards are taken from the top of a well-shuffled deck of cards. What is the probability that they are all black? Similarly, find the probability that there are 2, 1, or no blacks in the selection.

A natural sample space to consider is the set S of triples (x, y, z), where x, y, z are different cards. Using the multiplication principle, there are $52 \cdot 51 \cdot 50$ possibilities. To find the probability that all cards are black, we must find the number of triples in S in which all components are black. Using the multiplication principle again, there are $26 \cdot 25 \cdot 24$ such possibilities. Hence

$$p(\text{all cards are black}) = \frac{26 \cdot 25 \cdot 24}{52 \cdot 51 \cdot 50} = \frac{2}{17} (= .118)$$

To find the probability that two of the cards are black, we consider 3 cases: The red card is the first, second, or third card chosen. (This is an "it depends" situation.) If the red card is the first card, the multiplication principle shows that there are $26 \cdot 26 \cdot 25$ possibilities. The *same* answer is found in the other 2 cases. Thus there is a total of $3 \cdot 26 \cdot 26 \cdot 25$ possibilities with 1 red and 2 black cards, and

$$p(\text{exactly two blacks}) = \frac{3 \cdot 26 \cdot 26 \cdot 25}{52 \cdot 51 \cdot 50} = \frac{13}{34} (= .382)$$

The number of possibilities for 2 black cards can also be computed by the following 4-stage scheme (posed as questions): 1. Where is the red card? 2. What red card appears? 3. What is the first black card that appears?

4. What is the next black card? In this manner we may obtain $3 \cdot 26 \cdot 26 \cdot 25$ possibilities directly.

The probabilities for only 1 black card, and for no black cards, need not be computed anew. These cases are the same as the cases "2 red" and "all red," and these probabilities have already been found, if we reverse the roles of red and black.

We summarize the results in Table 2.9. The reader should compare with Exercise 4, Section 1 of Chapter 1.

2.9   Count of Black Cards Among 3 Cards Chosen at Random

Number of black cards	0	1	2	3
Probability	$\frac{2}{17}$	$\frac{13}{34}$	$\frac{13}{34}$	$\frac{2}{17}$
Probability (percent)	11.8	38.2	38.2	11.8

16   Example

Five men and 5 women are invited to a dinner party. They quickly notice that the sexes alternate about the table. (Each person was assigned to a specific seat. The table had 10 places.) When it was suggested that this was no accident, the hostess claimed that it was in fact due to chance. Was she telling the truth?

Here, there are 10 chairs, which we think of as the stages. We want to find the probability of alternating sexes. Using the multiplication principle, the sample space—all the possible arrangements—has $10 \cdot 9 \cdot 8 \cdot 7 \cdot 6 \cdot 5 \cdot 4 \cdot 3 \cdot 2 \cdot 1$ elements. The number of arrangements in which the sexes alternate about the table is found by the multiplication principle. Going around the table from a fixed seat, we find that there are $10 \cdot 5 \cdot 4 \cdot 4 \cdot 3 \cdot 3 \cdot 2 \cdot 2 \cdot 1 \cdot 1$ possibilities. The probability is

$$p = \frac{10 \cdot 5 \cdot 4 \cdot 4 \cdot 3 \cdot 3 \cdot 2 \cdot 2 \cdot 1 \cdot 1}{10 \cdot 9 \cdot 8 \cdot 7 \cdot 6 \cdot 5 \cdot 4 \cdot 3 \cdot 2 \cdot 1} = \frac{1}{126} = .0079$$

In all probability, this sort of an arrangement was no accident.

EXERCISES

1. A drawer contains 4 black and 4 red socks. Neil grabs 2 socks at random. What is the probability that they match?

2. How many 3-digit numbers are there? Of these, how many are there which
   a. begin with the digit 2, 3, or 4?

  b. begin with the digit 2, 3, or 4 and have no repeated digits?
  c. have 1 digit repeated?
  d. do not contain the digit 5?
  e. contain the digit 5 and have no repeated digits?
  f. contain the digits 5 and 9?
  g. do not contain either of the digits 5 or 9?
  h. contain the digit 5 or 9 but not both?

3. A box of diaper pins contains 20 good pins and 3 defective ones. Three diaper pins are chosen at random from the box. What is the probability that they are all good?

4. Three cards are chosen at random from a deck of cards. What is the probability that none are aces? What is the probability that none are picture cards (jack, queen, or king)?

5. Four people meet at a party and are rather surprised to discover that 2 of them were born in the same month. Find the probability that 4 people chosen at random were born in different months. (You may assume, as a good approximation, that the months are equally likely.)

6. In how many ways can 3 people sit at a lunch counter with 10 seats? How many ways can they sit so that no 2 of the people sit next to one another?

7. Five dice are tossed. What is the probability that all 5 dice will turn up different numbers?

8. Three cards are chosen at random from a standard deck. Find the probability of choosing
  a. different suits.
  b. all the same suit.
  c. different ranks.

9. (*Three-card poker*) Three cards are chosen at random from a standard deck. Find the probability of choosing
  a. a royal flush (Q, K, A all of the same suit).
  b. a straight flush (3 consecutive ranks, all of the same suit; we treat an an ace as a 1, and the jack, queen, and king as 11, 12, and 13, respectively).
  c. a flush (all cards of the same suit but not a royal flush or a straight flush).
  d. a straight (3 consecutive cards not all of the same suit; here the convention is that an ace may be regarded as either a 1 or a 14, and the convention for the picture cards is as in part b).
  e. 3 of a kind (all cards of the same rank).

    f. a pair (2 cards of the same rank and 1 of a different rank).

    g. a bust (none of the above).

10. An urn contains 7 red balls and 3 green ones. If 3 different balls are chosen from the urn, what is the probability that they are all red?

11. In Example 15 the probability p_3 of choosing 3 black cards was shown to be $\frac{2}{17}$. For reasons of symmetry, we noted that $p_2 = p_1$ and $p_3 = p_0$. (Here p_i is the probability of finding i black cards.) Using these facts alone, find p_2. (*Hint:* Use Equation 1.5, after you have identified the appropriate sample space.)

12. Three people each choose a number from 1 to 10 at random. What is the probability that at least 2 of the people chose the same number?

13. A president, vice president, and treasurer are to be chosen from a club that has 12 people $(A, B, C, D, E, F, G, H, I, J, K, L)$. How many ways are there to choose the officers if

    a. A and B will not serve together.

    b. C and D will either serve together or not at all.

    c. E must be an officer.

    d. F is either president or will not serve.

    e. I and J must be officers.

    f. K must be an officer and he must serve with L and/or G.

3   PERMUTATIONS AND COMBINATIONS

There is a joke that goes as follows. Gus: "There are 100 cows in that field." Bob: "How do you know?" Gus: "I counted 400 feet and divided by 4." Anyone who has ever tried to explain a joke knows that it is not worth doing. Nevertheless, the mathematics can be explained simply enough. Gus counted 400 legs. Hence each cow was counted 4 times. Thus there were $400/4 = 100$ cows. This illustrates the division principle. The strange feature is that it is actually useful. It is sometimes easier to count legs than cows, even if there are more legs.

17 Theorem. The Division Principle

Let S be a set with n elements, and suppose that each element of S determines an element of a set T. Suppose further that each element of T is determined by exactly r elements of S. Then there are $k = n/r$ elements of T.

In this theorem we think of using S to count the elements of T. The hypothesis tells us that each element of T is counted r times.

In the joke above, S was the set of legs ($n = 400$). Each leg determined a cow (an element of T, the set of cows). Each cow was determined by exactly 4 legs ($r = 4$). Therefore, there were $k = \frac{400}{4} = 100$ cows.

To prove Theorem 17, suppose the elements of T are $t_1, t_2, \ldots, t_k$. Let A_i be the set of elements in S which determine t_i. By hypothesis, A_i has r elements. But every element of S is in exactly one of the A_i's. Therefore, the number of elements in S is the sum of the number elements in the A_i's. Thus $n = r + \cdots + r = kr$. Hence $k = n/r$, which is the result.

18 Example

In a set containing 10 elements, how many subsets are there which contain 3 elements?

Suppose the set is $A = \{1, 2, \ldots, 10\}$. We may form a subset of 3 elements by first choosing any number in A (10 possibilities), then choosing another in A (9 possibilities), and finally choosing the last number (8 possibilities). There are $10 \cdot 9 \cdot 8$ ways of doing this. This ordered triple determines a subset of 3 elements. But a subset, such as $\{2, 5, 7\}$, is determined by several different ordered triples: $(2, 5, 7)$, $(5, 7, 2)$, etc. In fact, we can see that any subset is determined by $3 \cdot 2 \cdot 1$ ordered triples. This is precisely the situation of Theorem 17. The set S of ordered triples has $n = 10 \cdot 9 \cdot 8$ elements, and the set T consists of subsets with 3 elements. Each element of S determines a subset (an element of T), and each subset is determined by $r = 3 \cdot 2 \cdot 1$ element of S. Thus there are $k = 10 \cdot 9 \cdot 8/3 \cdot 2 \cdot 1 = 120$ subsets with 3 elements.

This is a typical use of the division principle. We first count, using the multiplication principle. Then we consider how many repetitions of each case occur and divide by that number.

To generalize this example, it is convenient to use the language of *sampling*. Sampling occurs when we have a population and we wish to choose some members of this population in order to observe some of their characteristics. Some examples are as follows.

1. A polltaker wants to see how the people in a large city feel about an issue. Usually, only relatively few people (the sample) are interviewed.

2. A quality-control expert might want to find out how close to specifications a factory-produced transistor is. He has a pile of 5,000 transistors. He chooses some of these (his sample) and tests them.

3. When we throw 5 dice, we have a sample of size 5 from the population $\{1, 2, 3, 4, 5, 6\}$.

4. A poker hand (5 cards) is a sample of size 5 from a population of 52.

5. A class of 25 in a school whose enrollment is 600 may be considered a sample.

The general situation may be described as follows. We start with a *population*, which is a set P. We then choose a *sample of size* r. There are two broad categories: (1) ordered or unordered, and (2) with or without replacement. An *ordered sample* of size r is merely an ordered r-tuple $(a_1, \ldots, a_r)$ of elements of P. We have an *unordered sample* if we agree to regard 2 r-tuples as identical if one can be reordered into the other. In the 5-coin example, HHTHT and THHTH are different ordered samples of size 5 (from the population $\{H, T\}$) but are regarded as identical unordered samples. If we pick up 5 cards to form a poker hand, it is customary to regard this hand as an unordered sample. We usually do not care which card was picked up first, second, etc.

A sample $(a_1, \ldots, a_r)$ *without replacement* occurs when the elements a_i are distinct. We imagine an urn filled with marbles and we pick r different marbles. Similarly, when we choose 5 cards, they are different. In a sampling *with replacement*, repetitions are allowed. In an urn situation, after we pick a marble, we put it back before we pick another. A policeman who tickets motorists (the population) does it in order (his tickets are numbered), and he is not prohibited from ticketing the same fellow twice (a sampling with replacement).

In many cases the decision as to whether a sampling is to be regarded as ordered or not, or even with or without replacement, is somewhat arbitrary and depends upon the intended use. For example, in the 5-coin example (Example 1.16) we decided on an ordered sample because it seemed reasonable that these were equally likely.

Using the language of sampling, Example 18 called for the number of *unordered samples* of size 3 *without replacement* from a population of size 10. To generalize this example, we introduce the following definition.

19 Definition

$$n! = 1 \cdot 2 \cdot 3 \cdot \ldots \cdot n \tag{2.5}$$

The notation $n!$ is read "n factorial." The definition does not imply that n must be larger than 3. Rather, starting at 1, we keep multiplying successive numbers until we reach n. Equivalently, $n! = n(n-1) \cdots 1$. Thus $1! = 1, 2! = 2 \cdot 1 = 2, 3! = 3 \cdot 2 \cdot 1 = 1 \cdot 2 \cdot 3 = 6$, etc. Table 2.10 is a brief table of factorials. It is seen that $n!$ grows very fast with n. For example, $52! = 8.066 \times 10^{67}$ to 4 significant figures. This is the number of possible ways of arranging a pack of cards.

2.10 Values of $n!$, $1 \leq n \leq 10$

$1! =$	1	$6! =$	720
$2! =$	2	$7! =$	5,040
$3! =$	6	$8! =$	40,320
$4! =$	24	$9! =$	362,880
$5! =$	120	$10! =$	3,628,800

20   Theorem

If S has n elements, there are $n!$ ways of arranging S in order. Equivalently, there are $n!$ ordered samples of size n without replacement from a population of size n.

This theorem is an immediate consequence of the multiplication principle and needs no further proof.

An *ordered* sample of size r without replacement is also called a *permutation* of size r. Thus we may say that there are $n!$ permutations of size n using a population of size n. If smaller samples are considered, we use the following definition and theorem.

21   Definition

If $r \leqslant n$,

$$_nP_r = \underbrace{n(n-1) \cdots (n-r+1)}_{r \text{ factors}} \tag{2.6}$$

Thus $_{10}P_3 = 10 \cdot 9 \cdot 8$. In the symbol $_nP_r$, n is used to start the multiplication, and we successively reduce each factor by 1. The number of factors is r. Thus $_{10}P_3 = 10 \cdot 9 \cdot 8$ and $_{50}P_{17} = 50 \cdot 49 \cdots 34$. Here the factor 34 was computed as in Equation 2.6: $50 - 17 + 1 = 34$. The letter P is used to recall the word "permutation."

22   Theorem

If S has n elements, the number of ordered samples of size r without replacement is $_nP_r$. Equivalently, the number of permutations of size r from a population of size n is $_nP_r$.

Theorem 22 is also an immediate consequence of the multiplication principle. In fact, Theorem 20 is a special case of Theorem 22, using

$$n! = {_nP_n} \tag{2.7}$$

This equation is an immediate consequence of Equation 2.6 and Definition 19.

An alternative form for expressing $_nP_r$ is obtained by multiplying the numerator and denominator of the right-hand side of Equation 2.6 by $(n-r)!$. For example,

$$_{10}P_3 = 10 \cdot 9 \cdot 8 = \frac{10 \cdot 9 \cdot 8 \cdot 7 \cdot 6 \cdot 5 \cdot 4 \cdot 3 \cdot 2 \cdot 1}{7 \cdot 6 \cdot 5 \cdot 4 \cdot 3 \cdot 2 \cdot 1} = \frac{10!}{7!}$$

In general,

$$_nP_r = \frac{n!}{(n-r)!} \tag{2.8}$$

A brief examination of Table 2.10 shows that Equation 2.8 is useful for theoretical purposes only. We do not calculate $_{10}P_2$ by dividing 10! by 8!.

An *unordered* sample of size r without replacement is called a *combination* of size r. Equivalently, a combination of size r is the same as a set with r elements. We may generalize Example 18 with the following definition and theorem.

23 Definition

If $1 \leqslant r \leqslant n$,

$$_nC_r = \binom{n}{r} = \frac{_nP_r}{r!} = \frac{n(n-1) \cdots (n-r+1)}{r(r-1) \cdots 1} \tag{2.9}$$

Here the symbol $_nC_r$ is used in analogy with $_nP_r$. The letter C indicates that combinations are being found. However, the symbol $\binom{n}{r}$ is now in rather general use and we shall generally use it in this text. Note that $\binom{n}{r}$ is not the fraction $\frac{n}{r}$. There is no horizontal bar, and parentheses must be used. For example,

$$\binom{10}{3} = \frac{10 \cdot 9 \cdot 8}{3 \cdot 2 \cdot 1} = 120 \qquad \binom{n}{2} = \frac{n(n-1)}{2 \cdot 1} \qquad \binom{n}{1} = \frac{n}{1} = n$$

24 Theorem

If S has n elements, the number of unordered samples of size r without replacement is $\binom{n}{r}$. Equivalently, there are $\binom{n}{r}$ subsets of S with r elements.

The proof of this theorem is exactly as in Example 18, and we spare the reader the details. We note here that it is not even immediately evident from Equation 2.9 that $\binom{n}{r}$ is an integer. Thus it may appear to be a coincidence that $\frac{11 \cdot 10 \cdot 9 \cdot 8}{4 \cdot 3 \cdot 2 \cdot 1}$ has enough cancellations to be a whole number. But this is clear from Theorem 24.

If we use Equations 2.8 and 2.9, we derive an alternative formula for $\binom{n}{r}$. Thus

$$\binom{n}{r} = \frac{_nP_r}{r!} = \frac{n!}{(n-r)!\,r!}$$

and we have

$$\binom{n}{r} = \frac{n!}{r!(n-r)!} \tag{2.10}$$

For example, $\binom{10}{3} = 10!/3!7!$. By the same formula, however, $\binom{10}{7} =$ $10!/7!3!$. Thus $\binom{10}{7} = \binom{10}{3}$. In general we have

$$\binom{n}{r} = \binom{n}{n-r}$$

(2.11)

To prove this we use Equation 2.10. We have

$$\binom{n}{n-r} = \frac{n!}{(n-r)!(n-(n-r))!} = \frac{n!}{(n-r)!\,r!} = \binom{n}{r}$$

which completes the proof.

Equation 2.11 may also be proved in the following way. To choose a subset of r from a set of n things, it suffices to choose the $(n-r)$ objects that are not to be in the set. Thus, to specify which 15 of 18 books are to be taken from a library, it is only necessary to specify which 3 are to be left behind. The number of possibilities is the same in either case.

It is convenient to define $0!$, $_nP_0$, and $\binom{n}{0}$. This is a convention, similar to the one used in algebra to define $a^0 = 1$. The decision is governed by an

2.11 $_nP_r, 0 \leqslant r \leqslant n \leqslant 8$

n\\r	0	1	2	3	4	5	6	7	8
0	1								
1	1	1							
2	1	2	2						
3	1	3	6	6					
4	1	4	12	24	24				
5	1	5	20	60	120	120			
6	1	6	30	120	360	720	720		
7	1	7	42	210	840	2,520	5,040	5,040	
8	1	8	56	336	1,680	6,720	20,160	40,320	40,320

attempt to make our formulas true for $r = 0$. We thus define

$$0! = 1 \qquad {}_nP_0 = 1 \qquad \binom{n}{0} = {}_nC_0 = 1 \qquad (2.12)$$

These definitions make Equations 2.7 through 2.11 true.

Tables 2.11 and 2.12 give brief tables of ${}_nP_r$ and $\binom{n}{r}$ for small values of n and r.

We conclude this section with an example that illustrates one use of combinations.

25 Example

Ten coins are tossed. What is the probability that 5 are heads and 5 are tails? Similarly, find the probability of r heads ($0 \leqslant r \leqslant 10$).

2.12 $\binom{n}{r}, 0 \leqslant r \leqslant n \leqslant 10$

n \ r	0	1	2	3	4	5	6	7	8	9	10
0	1										
1	1	1									
2	1	2	1								
3	1	3	3	1							
4	1	4	6	4	1						
5	1	5	10	10	5	1					
6	1	6	15	20	15	6	1				
7	1	7	21	35	35	21	7	1			
8	1	8	28	56	70	56	28	8	1		
9	1	9	36	84	126	126	84	36	9	1	
10	1	10	45	120	210	252	210	120	45	10	1

Let $A = \{H, T\}$. We take the sample space to be $A \times A \times \cdots \times A$ (10 factors). Equivalently, a sample point is an ordered sample of size 10 with replacement from the set A. There are $2^{10} = 1,024$ sample points, which we take to be equally likely. We wish to count how many have 5 heads. A sample point with 5 heads is specified by determining *where* those heads occur. Thus any subset of size 5 of $\{1, 2, \ldots, 10\}$ determines a sample with exactly 5 heads if we think of the numbers as determining a location. For example, the subset $\{2, 4, 5, 7, 10\}$ determines the sample point THTHHT-HTTH, and conversely. Since there are $\binom{10}{5}$ subsets of size 5 from a population of size 10, we have exactly $\binom{10}{5} = 252$ of the required sample points. The probability is

$$p(5H) = \frac{252}{1,024} \ (= 24.6\%)$$

In the same manner, we may find that the probability of r heads is $_{10}C_r/2^{10}$, $r = 0, 1, \ldots, 10$. The results are summarized in Table 2.13. The probabilities

2.13   Number of Heads Among 10 Coins

Number of heads	0	1	2	3	4	5	6	7	8	9	10
Probability	$\frac{1}{1024}$	$\frac{10}{1024}$	$\frac{45}{1024}$	$\frac{120}{1024}$	$\frac{210}{1024}$	$\frac{252}{1024}$	$\frac{210}{1024}$	$\frac{120}{1024}$	$\frac{45}{1024}$	$\frac{10}{1024}$	$\frac{1}{1024}$
Probability (percent)	0.1	1.0	4.4	11.7	20.5	24.6	20.5	11.7	4.4	1.0	0.1

are indicated graphically in Fig. 2.14. The reader should compare these probabilities with his experimental results for Exercise 7, Section 1 of Chapter 1.

EXERCISES

1. Evaluate, without using tables:
   a. $6!$
   b. $8!/5!$
   c. $_5P_2$
   d. $_{50}P_3$
   e. $_{12}C_3$
   f. $_{80}P_7/_{81}P_7$
   g. $_{50}C_{46}$
   h. $\binom{20}{4} \Big/ \binom{19}{3}$
   i. $(3!)!$
   j. $8!/(4!)^2$

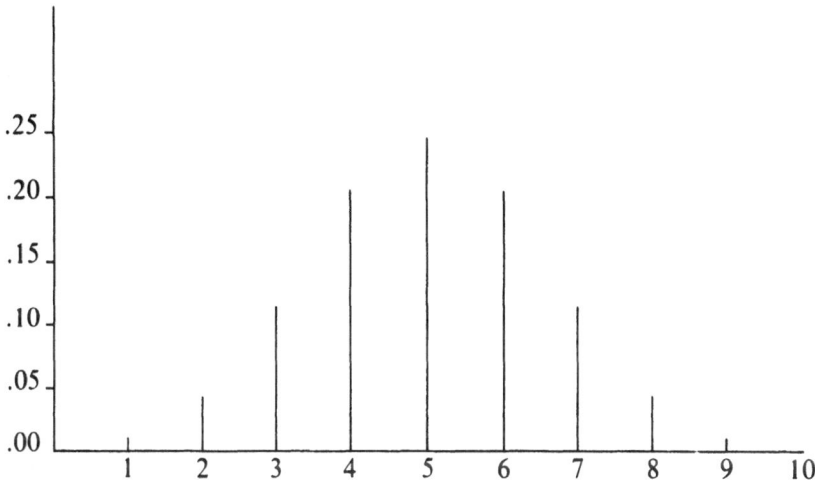

2.14   Number of Heads Among 10 Tossed Coins

2. Express each of the following using factorial notation.

a. $_{12}P_2$ b. $51 \cdot 50!$

c. $20 \cdot 19 \cdot 18!$ d. $_{20}C_{16}$

e. $n(n-1)$ f. $(n+2)(n+1)(n)$

g. $2 \cdot 4 \cdot 6 \cdot 8 \cdot 10 \cdots 48$ h. $1 \cdot 3 \cdot 5 \cdot 7 \cdots 21$

3. Add, expressing your answer in factorial notation, as far as is feasible.

a. $\dfrac{1}{10!} + \dfrac{1}{11!}$

b. $\dfrac{1}{7!9!} + \dfrac{1}{6!10!}$

c. $\dfrac{18!}{8!10!} + \dfrac{18!}{7!11!}$

4. Using Equation 2.10, prove that

$$\binom{n}{r} + \binom{n}{r+1} = \binom{n+1}{r+1}$$

5. From a club with 12 members, how many committees of 4 may be found?

6. There are 20 points on the circumference of a circle. It is desired to form triangles choosing vertices from these points. How many triangles can be formed?

7. In Exercise 6 it is decided that a given point must be used as a vertex for the triangle. How many triangles are possible with this additional restriction?

8. The figure represents the layout of streets in a certain city. A man starts at *A* and walks along the streets to *B* always moving north or east.

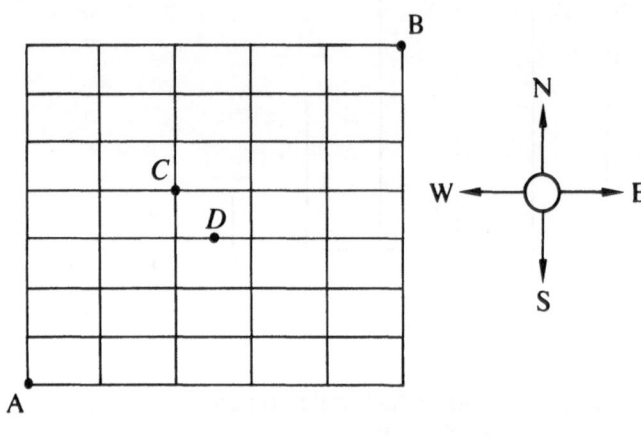

A Walk

    a. How many paths is it possible for him to take?
    b. If he insists on passing by the intersection *C* to view a certain billboard, how many paths from *A* to *B* are available to him?
    c. If he must pass the candy store located at *D*, how many paths (from *A* to *B*) are available to him?

9. Each of the following may be regarded as a sample taken from a population. Give the population and the size of the sample, if possible. State whether the sample is ordered or unordered and whether it is with replacement or without replacement. In some cases there are several possible answers. If so, give the possibilities and explain the ambiguity.
    a. A word in the English language.
    b. A 4-digit number.
    c. A poker hand (5 cards).
    d. The result of tossing 8 dice.
    e. The result of tossing 10 coins.
    f. A listing of the top 5 students, in order of ranking, from a graduating class.
    g. A baseball team.
    h. The final standings in the American League.
    i. The first-division teams in the National League.

In Exercises 10 *through* 14, *express your answer using the* $_nP_r$, $\binom{n}{r}$, *or n! notation. It is not necessary to evaluate your answer.*

10. If 3 cards are chosen from a deck of cards, what is the probability they are all black?

11. How many possible poker hands (5 cards) are there if we do not count the order in which the cards are picked up? What is the probability that a poker hand consists only of spades?

12. A certain experimenter wishes to discover the eating habits of hamsters. He is sent 20 hamsters, of which 2 happen to have enormous appetites. He chooses 5 of the hamsters at random. What is the probability that he does not choose the big eaters?

13. A man has 30 books, but only 20 can fit on his bookshelf. How many ways can he arrange his bookshelf?

14. An urn contains 6 white balls and 9 black balls. Two balls are chosen at random. What is the probability that they are both black? (Do this problem using ordered as well as unordered samples.)

15. Any ordered n-tuple determines an ordered r-tuple; namely, its first r components. Using this fact, prove Equation 2.8 using the division principle and Theorem 20. Similarly, prove Equation 2.10 using the division principle, the multiplication principle, and Theorem 20.

4 BINOMIAL COEFFICIENTS

A brief inspection of Table 2.12 shows an interesting and useful property of the numbers $\binom{n}{r}$. If 2 consecutive entries on one row are added together, the sum is the entry on the next row, under the second summand. Referring to table 2.12, row $n = 6$, we see under columns $r = 2$ and 3 that $15 + 20 = 35$, the entry in row $n = 7$ and column 3. Thus $\binom{7}{3} = \binom{6}{2} + \binom{6}{3}$. We shall shortly prove this in general, but let us now note that this fact gives an excellent technique for constructing this table. For example, starting with the row $n = 3$ (1, 3, 3, 1), we can construct the next row $n = 4$ by putting 1 at each end and adding as indicated in Fig. 2.15. This process can be continued indefinitely and in this way we can construct a table of $\binom{n}{r}$ without using the

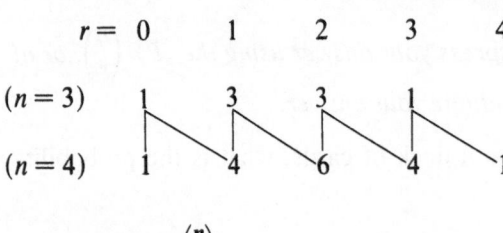

$$r = \quad 0 \qquad 1 \qquad 2 \qquad 3 \qquad 4$$

$(n = 3) \qquad 1 \qquad 3 \qquad 3 \qquad 1$

$(n = 4) \qquad 1 \qquad 4 \qquad 6 \qquad 4 \qquad 1$

2.15 Property of $\binom{n}{r}$

definition of $\binom{n}{r}$ given by Equation 2.9. In fact, the whole process can be started with the single entry 1 corresponding to $n = 0$, $r = 0$. [The more conservative reader might prefer the row $(1, 1)$, corresponding to $n = 1$.] Table 2.12 is called *Pascal's triangle*, in honor of the great seventeenth-century mathematician. We shall refer to the method illustrated in Fig. 2.15 as the Pascal method for generating the numbers $\binom{n}{r}$. We now prove the general result.

26 Theorem

$$\binom{n+1}{r+1} = \binom{n}{r} + \binom{n}{r+1} \tag{2.13}$$

This equation can be proved directly using Equation 2.10 and some algebraic manipulation. (This is Exercise 4 of Section 3.) We shall give an alternative proof using Theorem 24. Suppose S is a set with $n+1$ elements and we wish to form a subset with $r+1$ elements. There are $\binom{n+1}{r+1}$ such subsets. Now fix element a_1 in S, and distinguish between the subsets that contain a_1 or do not. If a_1 is to be in a subset, there are only r choices from the remaining n elements of S, hence there are $\binom{n}{r}$ of the subsets with a_1 in them. If a_1 is not to be in a subset, we must choose $r+1$ elements from the remaining n elements of S. There are $\binom{n}{r+1}$ ways of doing this. Summarizing, we have

1. $\binom{n}{r}$ subsets of size $r+1$ with a_1 in them.

2. $\binom{n}{r+1}$ subsets of size $r+1$ with a_1 not in them.

Thus there are $\binom{n}{r} + \binom{n}{r+1}$ subsets of size $r+1$ in all. But we also know that there are $\binom{n+1}{r+1}$ subsets of size $r+1$ in all. Hence we have the result.

Using Equation 2.13 we can prove the famous binomial theorem.

27 Theorem. The Binomial Theorem

$$(1+x)^n = 1+\binom{n}{1}x+\binom{n}{2}x^2+\cdots+\binom{n}{n-1}x^{n-1}+x^n$$

$$= \sum_{r=0}^{n} \binom{n}{r}x^r \tag{2.14}$$

For example, corresponding to row 5 of Table 2.12 we have

$$(1+x)^5 = 1+5x+10x^2+10x^3+5x^4+x^5$$

To prove Formula 2.14 in general, we start with a known case, say $n = 2$. We have $(1+x)^2 = 1+2x+x^2$. Multiply by $(1+x)$ to obtain $(1+x)^3 = (1+2x+x^2)(1+x)$. This multiplication may be performed as follows:

$$
\begin{array}{r}
1+2x \ +x^2 \\
1 \ +x \\
\hline
x+2x^2+x^3 \\
1+2x+ \ x^2 \\
\hline
1+3x+3x^2+x^3
\end{array}
$$

Thus $(1+x)^3 = 1+3x+3x^2+x^3$, which is Equation 2.14 for $n = 3$. Multiplying again by $(1+x)$, we have $(1+x)^4 = (1+3x+3x^2+x^3)(1+x)$. Let us multiply as above, using the coefficients only.

$$
\begin{array}{l}
1+3x+3x^2+x^3 \quad \rightarrow \\
 1+x \\
\end{array}
\qquad
\begin{array}{r}
1 \ 3 \ 3 \ 1 \\
1 \ 1 \\
\hline
1 \ 3 \ 3 \ 1 \\
1 \ 3 \ 3 \ 1 \\
\hline
1 \ 4 \ 6 \ 4 \ 1
\end{array}
$$

Thus $(1+x)^4 = 1+4x+6x^2+4x^3+x^4$. These examples show (and it is easy to see in general) that to multiply by $1+x$, we slide the coefficients one over to the left, and add. The results are the new coefficients, starting with the constant term and proceeding through increasing powers of x. Therefore, this step-by-step method of constructing the coefficients of $(1+x)^n$ is identical to Pascal's method of constructing the numbers $\binom{n}{r}$. Since they both start with $(1, 2, 1)$ for $n = 2$, or $(1, 1)$ for $n = 1$, and since the method gives a unique answer for each n, the coefficients in $(1+x)^n$ are identical to the numbers $\binom{n}{r}$ $(r = 0, 1, \ldots, n)$. This completes the proof.

Because of Theorem 27, the numbers $\binom{n}{r}$ are also called the *binomial coefficients*.

A more general form for Equation 2.14 involves $(x+y)^n$. Thus

$$(x+y)^n = x^n + \binom{n}{1}x^{n-1}y + \cdots + \binom{n}{n-1}xy^{n-1} + y^n$$

$$= \sum_{r=0}^{n} \binom{n}{r}x^{n-r}y^r \qquad (2.15)$$

This follows directly from Equation 2.14 by substituting y/x for x in that formula and simplifying. Thus, from Equation 2.14, we obtain

$$\left(1+\frac{y}{x}\right)^n = 1 + \binom{n}{1}\frac{y}{x} + \cdots + \left(\frac{y}{x}\right)^n$$

Multiplying by x^n we have

$$x^n\left(1+\frac{y}{x}\right)^n = x^n + \binom{n}{1}x^{n-1}y + \cdots + y^n$$

Finally, the left-hand side may be simplified to $(x+y)^n$:

$$x^n\left(1+\frac{y}{x}\right)^n = \left[x\left(1+\frac{y}{x}\right)\right]^n = (x+y)^n$$

This proves Equation 2.15. This equation is also called the binomial theorem.

We may give the following alternative proof of Equation 2.14. In the expansion $(1+x)^n = (1+x)(1+x)\cdots(1+x)$ we multiply out as follows. Choose either a 1 or an x in each factor $(1+x)$ and multiply. Then add the results. The coefficient of x^r is precisely the number of ways we can choose r x's from among these n factors. By Theorem 24 this number is $\binom{n}{r}$.

Q.E.D.

If we substitute $x = 1$ into Equation 2.14, we obtain

$$2^n = \binom{n}{0} + \binom{n}{1} + \binom{n}{2} + \cdots + \binom{n}{n-1} + \binom{n}{n} \qquad (2.16)$$

This equation is not surprising. By Example 10 we know that there are 2^n subsets of a set with n elements, $\binom{n}{1}$ have 1 element, $\binom{n}{2}$ have 2 elements, etc. (Theorem 24). Thus Equation 2.16 may be regarded as a way of counting all the subsets of a set with n elements.

If we substitute $x = -1$ into Equation 2.14, we obtain

$$0 = \binom{n}{0} - \binom{n}{1} + \binom{n}{2} - \binom{n}{3} + \cdots \pm \binom{n}{n} \qquad (n \geq 1) \qquad (2.17)$$

Transposing the negative summands, we obtain

$$\binom{n}{1}+\binom{n}{3}+\cdots=\binom{n}{0}+\binom{n}{2}+\cdots \tag{2.18}$$

Here the sums continue on each side until the last term is either $\binom{n}{n}$ or $\binom{n}{n-1}$. For example,

$$5+10+1=1+10+5 \qquad (n=5)$$
$$6+20+6=1+15+15+1 \qquad (n=6)$$

Equation 2.18 may be interpreted: "If a set S is nonempty with finitely many elements, there are as many subsets with an even number of elements as with an odd number." In the language of probability, if a subset is chosen at random from a nonempty finite set, it will have an even number of elements with probability $\frac{1}{2}$.

An important application of Equation 2.14 is to approximate powers of numbers that are near 1. Thus to compute $(1.03)^8$ we may use this equation to find

$$(1.03)^8 = (1+.03)^8 = 1+\binom{8}{1}(.03)+\binom{8}{2}(03)^2+\cdots$$
$$= 1+(8)(.03)+(28)(.0009)+(56)(.000027)+\cdots$$
$$= 1+.240+.0252+.001512+\cdots$$
$$= 1.267 \text{ (to 3 decimal places)}$$

This method is useful for small x, provided nx is also small. For in that case, the higher powers of x in Equation 2.14 rapidly become negligible, while the coefficients $\binom{n}{1}$, $\binom{n}{2}$, etc., are not large enough to overtake the powers of x. We may also use this technique for computing powers of numbers less than 1. Here it is useful to use a variant of Equation 2.14. If we replace x by $-x$, then we obtain

$$[1+(-x)]^n = 1+\binom{n}{1}(-x)+\binom{n}{2}(-x)^2+\cdots+(-x)^n$$

or

$$(1-x)^n = 1-\binom{n}{1}x+\binom{n}{2}x^2-\binom{n}{3}x^3+\cdots+(-1)^n x^n \tag{2.19}$$

For example,

$$(.98)^{10} = (1-.02)^{10} = 1-\binom{10}{1}(.02)+\binom{10}{2}(.02)^2-\binom{10}{3}(.02)^3+\cdots$$
$$= 1-(10)(.02)+(45)(.0004)-(120)(.000008)+\cdots$$
$$= 1-.2+.018-.00096+\cdots$$
$$= .817 \text{ (to 3 decimal places)}$$

These examples show that Equations 2.14 and 2.19 are written with the most important terms first, assuming x and nx are small. In many cases, depending on the accuracy required, it is not necessary to write out all the terms. In Section 1 of Chapter 4 we shall see how to deal with a case such as $(1.002)^{2,000}$, in which $x = .002$ is small but $nx = 4$ is moderate in size.

The following example gives another, somewhat surprising, application of the binomial coefficients.

*28 Example

Eight dice are tossed. If the dice are identical in appearance, how many different-looking (distinguishable) occurrences are there?

What is called for is the number of *unordered samples* of size 8 *with replacement* from a population of size 6. An "appearance" is completely and uniquely specified by the number r_1 of 1's, r_2 of 2's, etc. Clearly $r_1 + r_2 + \cdots + r_6 = 8$, and $r_i \geqslant 0$. We use the following trick. We think of 8 balls on a line and 5 dividers. In the following diagram, the balls are the O's and dividers are the ×'s:

$$O \times OO \times \times OOO \times O \times O$$
$$1 \quad 2 \quad 0 \quad 3 \quad\; 1 \quad 1$$

The 5 dividers break up the balls into 6 (ordered) groups, and the number in the first group is taken to be r_1, etc. The above diagram has $r_1 = 1$, $r_2 = 2$, $r_3 = 0$, $r_4 = 3$, $r_5 = 1$, $r_6 = 1$. Clearly, any sequence of 5 ×'s and 8 O's determines the numbers r_i, and conversely. But the number of such sequences is $\binom{13}{5} = 1{,}287$, because we must choose 5 places from among 13 to locate the ×'s. Thus there are 1,287 different-looking throws of 8 dice.

We generalize this result in the following theorem.

*29 Theorem

If r is a nonnegative integer, the number of n-tuples $(r_1, r_2, \ldots, r_n)$ (each r_i is an integer) satisfying the conditions

(a) $$r_i \geqslant 0 \qquad (i = 1, \ldots, n)$$

(b) $$r_1 + r_2 + \cdots + r_n = r$$

is

$$\binom{n + r - 1}{n - 1}$$

In the above example we had $n = 6$, $r = 8$. The proof of this theorem is the same as that given in that example, and we omit it. As in the example, we have the following alternative formulation.

*29' Theorem

The number of unordered samples of size r, with replacement, from a population of size n is $\binom{n+r-1}{n-1}$.

In contrast with this result, we take note that the number of *ordered* samples of size r, with replacement, from a population of size n, is simply n^r.

EXERCISES

1. Simplify, using Equation 2.13,

a. $\binom{13}{5} + \binom{13}{6}$

b. $\binom{11}{5} + \binom{11}{6} + \binom{12}{7}$

c. $\binom{n}{r-1} + \binom{n}{r}$

d. $\binom{x+2}{r+3} + \binom{x+2}{r+4}$

2. Using the Pascal property twice, express $\binom{n+2}{r+2}$ as a sum involving terms of the form $\binom{n}{x}$. (*Hint:* See Fig. 2.16.)

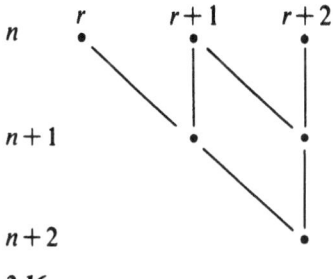

n r $r+1$ $r+2$

$n+1$

$n+2$

2.16

3. In analogy with the proof of Theorem 26, the number $\binom{n+2}{r+2}$ is the number of subsets of size $(r+2)$ of set S with $n+2$ elements. By singling out two elements a_1 and a_2 of S and distinguishing cases, prove[1]

$$\binom{n+2}{r+2} = \binom{n}{r} + 2\binom{n}{r+1} + \binom{n}{r+2}$$

---

[1] It is convenient to define $\binom{n}{r} = 0$ for $r > n$ and $r < 0$. With this definition, Theorem 24 is valid, and it is not necessary to restrict the size of r in equations such as this.

4. As in Exercise 3, prove that

$$\binom{n+3}{r+3} = \binom{n}{r} + 3\binom{n}{r+1} + 3\binom{n}{r+2} + \binom{n}{r+3}$$

5. Generalize Exercises 3 and 4 to prove that

$$\binom{n+b}{r+b} = \binom{n}{r} + \binom{b}{1}\binom{n}{r+1} + \binom{b}{2}\binom{n}{r+2} + \cdots + \binom{b}{b}\binom{n}{r+b} \qquad (2.20)$$

(*Hint:* It is convenient to think of a population of n pennies and b nickels. How many subsets of $r+b$ coins are there?)

6. Prove Equation 2.20 using the equation $(1+x)^{n+b} = (1+x)^n(1+x)^b$ and the binomial theorem. (*Hint:* Compare the coefficients of x^{r+b} on both sides of this equation.)

7. Prove that

$$\binom{2n}{n} = \binom{n}{0}^2 + \binom{n}{1}^2 + \cdots + \binom{n}{n}^2 \qquad (2.21)$$

as a special case of Equation 2.20.

8. Using the binomial theorem and Table 2.12, expand each of the following.

a. $(a+b)^5$ b. $(c-d)^7$
c. $(1+2y)^6$ d. $(2x+\tfrac{1}{2}y)^4$
e. $[x+(y+z)]^3$

9. Write out the first 3 terms in the binomial expansion of each of the following:

a. $(1+x)^{100}$ b. $(x-y)^{50}$
c. $(1-t)^{70}$ d. $(p+q)^n$

10. Using the identity $(1+x)^{2n}(1-x)^{2n} = (1-x^2)^{2n}$, prove that

$$\binom{2n}{0}^2 - \binom{2n}{1}^2 + \binom{2n}{2}^2 - \cdots + \binom{2n}{2n}^2 = (-1)^n\binom{2n}{n}$$

(*Hint:* Compare the coefficients of x^{2n} on both sides of this equation.)

11. Using the binomial expansion, evaluate each of the following numbers to 3 decimal places.

a. $(.97)^{10}$ b. $(1.0001)^{100}$
c. $(1+10^{-27})^{100,000}$ d. $(\tfrac{899}{900})^{30}$

12. A sample of size r is chosen from a population of size n. There are 4 possibilities if we classify with respect to ordering and with respect to replacement. State the number of samples possible in each of the 4 cases. Give reasons or quote theorems to justify your answer.

*13. An experiment has 4 possible outcomes. It is to be repeated 20 times and a frequency table is to be computed. How many possible tables are there? (It is not necessary to evaluate your answer.)

*14. How many different (unordered) coin combinations are there using 15 coins? (Only pennies, nickels, dimes, and quarters are to be used.)

*15. How many solutions of the equation $x_1 + x_2 + x_3 = 10$ are there? (Here x_i is a nonnegative integer.) How many solutions are there if each x_i is a positive integer? (*Hint:* Write $x_i = 1 + y_i$. Then y_i is nonnegative.) How many solutions are there with $x_1 \geq 2, x_2 \geq 1, x_3 \geq 1$?

*16. A teacher has a class of 10 students. It is grading time and he must assign each student a grade of A, B, C, D, or F.

    a. In how many different ways may he grade the class? (Do not evaluate.)

    b. The dean requires a grade distribution. How many possible grade distributions are there? (A grade distribution only tells how many A's, B's, etc., are given.)

    c. Suppose the teacher insists upon giving at least 1 A, at least 3 C's, and no F's. How many different grade distributions are available to him?

*17. Each student in a class of 10 is given an assignment to toss a die and record and result. The students are unreliable, and cannot be depended upon to carry out the assignment. A table is then prepared giving a frequency count for each of the 6 possible outcomes.

    a. By regarding this experiment as choosing an unordered sample of size 10 *or less*, without replacement, from a population of 6, show that the number of possible tables is

$$\binom{15}{5} + \binom{14}{5} + \binom{13}{5} + \cdots + \binom{5}{5}$$

    b. If a student does not do the experiment, he may report this fact; otherwise, he will report $1, 2, \ldots, 6$. Thus the population may be regarded as one of size 7. Using this idea, how many tables are possible?

    c. Combining parts a and b prove that

$$\binom{16}{6} = \binom{15}{5} + \binom{14}{5} + \cdots + \binom{5}{5}$$

*18. Generalize part b of Exercise 17 to compute the number of solutions of the inequality

$$r_1 + r_2 + \cdots + r_n \leq r$$

Here the r_i are taken to be nonnegative integers.

★19. Using the technique of Exercise 17, prove that

$$\binom{a}{a}+\binom{a+1}{a}+\cdots+\binom{a+n-1}{a}+\binom{a+n}{a}=\binom{a+n+1}{a+1} \quad (2.22)$$

Illustrate the result in Table 2.12. Write out Equation 2.22 for the special cases $a = 0$, $a = 1$, and $a = 2$.

5 APPLICATIONS OF THE MULTIPLICATION AND DIVISION PRINCIPLES

In Sections 2 and 3 we introduced the multiplication and division principles which led to the consideration of binomial coefficients. We now consider additional applications which use these principles in combination.

★30 Example

Ten dice are tossed. If we count the numbers 1 or 2 as low (L), 3 or 4 as middle (M), and 5 or 6 as high (H), what is the probability that 3 low, 3 middle, and 4 high numbers appear?

We note that this example is quite similar to Example 25 except that now we have 3 possibilities (L, M, or H) at each toss, rather than 2 (heads or tails). Let A be the set $\{L, M, H\}$. We choose as our sample space $A \times A \times \cdots \times A$ (10 factors). There are 3^{10} sample points, which we take as equally likely, because L, M, and H are all equally likely. To find the probability of 3 low, 3 middle, and 4 high, we must count the number of ordered 10-tuples that have 3 L's, 3 M's, and 4 H's appearing somewhere in them. There are $\binom{10}{3}$ ways to locate the L's, and after these are placed, we have $\binom{7}{3}$ ways of placing the M's. The remaining 4 places are then automatically filled with the H's. Thus there are $\binom{10}{3} \times \binom{7}{3}$ ways in which the result can occur. Using Equation 2.10, we may write

$$\binom{10}{3} \times \binom{7}{3} = \frac{10!}{3!7!}\frac{7!}{3!4!} = \frac{10!}{3!3!4!}$$

The required probability is therefore

$$\frac{10!}{3^{10}3!3!4!} \quad (= .071)$$

The form $10!/3!3!4!$ is most similar in appearance to the binomial coefficients $n!/r!(n-r)!$. The following results show how this example leads way to a generalization of the binomial coefficients.

*31 Definition

If an ordered sample of size r from the population $S = \{s_1, \ldots, s_n\}$ has r_1 of its components s_1, r_2 of its components s_2, etc., then we say that the *sample has type* $s_1^{r_1} \cdots s_n^{r_n}$.

Clearly $r_1 + \cdots + r_n = r$ and each $r_i \geqslant 0$. In the above example, $S = \{L, M, H\}$, $r = 10$, and we were finding the probability that a sample was of *type* $L^3 M^3 H^4$. Note that no multiplication is implied here (the "bases" s_i are not even necessarily numbers), but the notation is suggestive. For example, if $S = \{H, T\}$ (for coins), the ordered sample HHTHHHTH has type $H^5 T^2$. One simplification in notation is to omit zero exponents. Thus in the above dice example, we might consider the type $L^3 H^7$ rather than $L^3 H^7 M^0$.

Theorem 29 or 29' showed how many types there were. The following theorem shows us how many ordered samples there are of a given type.

*32 Theorem

Let $S = \{s_1, \ldots, s_n\}$, let r_i be a nonnegative integer for each i, and let $r_1 + \cdots + r_n = r$. Then the number N of ordered r-tuples of type $s_1^{r_1} \cdots s_n^{r_n}$ is

$$N = \frac{r!}{r_1! \cdots r_n!} \qquad (r_1 + \cdots + r_n = r) \qquad (2.23)$$

To prove this theorem, we proceed as in Example 30. There are $\binom{r}{r_1}$ ways of locating the s_1's. Then there are $\binom{r - r_1}{r_2}$ ways of locating the s_2's, etc. Thus, by the multiplication principle and Equation 2.10, there are

$$N = \binom{r}{r_1}\binom{r - r_1}{r_2}\binom{r - r_1 - r_2}{r_3} \cdots \binom{r - r_1 - \cdots - r_{n-1}}{r_n}$$

$$= \frac{r!}{r_1!(r - r_1)!} \cdot \frac{(r - r_1)!}{r_2!(r - r_1 - r_2)!} \cdots \frac{(r - r_1 - \cdots - r_{n-1})!}{r_n! 0!}$$

because $r - r_1 - \cdots - r_n = 0$. After extensive cancellations we have the result. (Another somewhat more direct proof is indicated in the exercises.)

Note that the number N of Equation 2.23 is a generalization of the binomial coefficients. (For the binomial coefficients we have $n = 2$.) The numbers N of this equation are called *multinomial (or k-nomial) coefficients*, for reasons to be seen below.

The binomial coefficients were used to count the number of subsets. This generalizes immediately to the following theorem, which is an alternative formulation of Theorem 32.

*32′ Theorem

Let L have r elements. The number of ways L can be partitioned into n subsets $L_1, \ldots, L_n$ which exhaust L, which have no elements in common, and which have $r_1, r_2, \ldots, r_n$ elements, respectively $(r_1 + \cdots + r_n = r)$, is given by Equation 2.23.

To see this, there are $\binom{r}{r_1}$ ways of choosing L_1, $\binom{r - r_1}{r_2}$ ways of choosing L_2, etc. Hence we have the same calculation as in Theorem 32. In that theorem we may regard L as the r locations in an ordered r-tuple, L_1 as the locations of the s_1's, etc. If we think of L as a population, then we may regard the L_i's as subpopulations. Theorem 32′ counts the number of ways of *splitting a population into subpopulations* of a different specified type. For example, if the teacher in Exercise 15 of Section 4 decides to assign 2 A's, 3 B's, and 5 C's to his class of 10, he has $10!/2!3!5!$ ways of making the grade assignments. We think of the class as the population, and we divide it into three subpopulations of the three specified types. Equivalently (if we order the students as in the rollbook), we are asking for grades of type $A^2B^3C^5$, and we may apply Theorem 32.

*33 Example

How many words may be made out of all the letters in the word "eerier"?

A word is understood to be an ordered sample of type $e^3 r^2 i^1$. Thus there are $6!/3!2!1! = 60$ words.

*34 Theorem. The Multinomial Theorem

$$(x_1 + x_2 + \cdots + x_k)^n = \sum_{r_1 + \cdots + r_k = n} \frac{n!}{r_1! \cdots r_k!} x_1^{r_1} \cdots x_k^{r_k} \qquad (2.24)$$

In this equation the sum is extended over all k-tuples $(r_1, \ldots, r_k)$ in which r_i is a nonnegative integer, and $r_1 + r_2 + \cdots + r_k = n$.

The proof proceeds by considering the product:

$$\underbrace{(x_1 + \cdots + x_k)(x_1 + \cdots + x_k) \cdots (x_1 + \cdots + x_k)}_{n \text{ factors}}$$

To find this product, we multiply any one of the summands x_i in the first factor by any summand in the second factor, etc., and add all the possibilities. By Theorem 32, a factor $x_1^{r_1} \cdots x_k^{r_k}$ occurs in exactly $n!/r_1! \cdots r_k!$ ways. This is the result.

*35 Example

Eighteen dice are thrown. What is the probability that each number shows up 3 times?

If we let $S = \{s_1, \ldots, s_6\}$ be the sample space for one die, then we choose S^{18} ($= S \times S \times \cdots \times S$ for 18 factors) as the required sample space. As usual we assume that S^{18} is uniform. We are looking for the probability that a sample point is of type $s_1{}^3 s_2{}^3 \cdots s_6{}^3$. The number of such sample points is $18!/(3!)^6$, by Theorem 32. The total number of sample points is 6^{18}. Therefore, the required probability is

$$\frac{18!}{(3!)^6} \cdot \frac{1}{6^{18}} = \frac{18!}{6^{24}} (= .00135)$$

The numbers in this calculation are so large that one might well ask how the answer is evaluated. One way is to use a table of logarithms. Most large handbooks of tables include a table of logarithms of factorials. Thus the logarithm of 18! can be read off directly, and the calculation to 3 decimal places is routine if one knows how to use logarithms.

Most people expect an answer much larger than .00137. This is why gamblers make money off the unsuspecting.

The following example generalizes Example 15.

36 Example

An urn contains 90 black balls and 40 red balls. Ten balls are selected at random (without replacement). What is the probability that 7 black and 3 red balls are chosen?

It is most convenient to use *unordered* samples. (Compare with Example 15 in which ordered samples were used.) We take as the sample space all unordered samples of 10 balls from among the 130. We regard this as a uniform space, because all possibilities seem equally likely. Thus there are $\binom{130}{10}$ outcomes.

We now compute the "favorable" outcomes. A favorable outcome is any set consisting of 7 black balls and 3 red balls. There are $\binom{90}{7}$ such sets of black balls and $\binom{40}{3}$ sets of red balls. Hence there are $\binom{90}{7}\binom{40}{3}$ favorable outcomes. Finally, the required probability is

$$p = \binom{90}{7}\binom{40}{3} \bigg/ \binom{130}{10} (= .277)$$

The best way to compute this answer is to use a table of logarithms and a table of logarithms of factorials. Thus, using Equation 2.11, we may express the answer in terms of factorials:

$$p = \frac{90!}{7!83!} \cdot \frac{40!}{3!37!} \cdot \frac{10!120!}{130!}$$

Then, by referring to a table of logarithms of factorials, this product can be easily evaluated. Incidentally, the numerator in the above expression for p is about 2.94×10^{391}. The answer, .277, is tame enough.

EXERCISES

*1. An urn has 3 balls in it, colored azure, beige, and chartreuse. Nine balls are chosen at random (with replacement) from this urn. What is the probability that 2 azure, 3 beige, and 4 chartreuse balls are chosen?

*2. In Exercise 1 find the probability that when 9 balls are chosen at random, 2 of one color, 3 of another, and 4 of the third color are chosen. Which is more likely—a 2-3-4 split or a 3-3-3 split? How much more likely is one than the other? (*Hint:* Divide your two answers to find out how much more likely one is than the other.)

*3. (*Alternative proof of Theorem 32'.*) If L is the set of r numbers $\{1, 2, \ldots, r\}$, then we may form a partition into subpopulations $L_1, L_2, \ldots, L_n$ (of size $r_1, \ldots, r_n$) by taking any arrangement of L, and then letting the first r_1 elements be the elements of L_1, the second r_2 elements the elements of L_2, etc. Using this idea and the division principle, prove Theorem 32'.

*4. Fifty people each shuffle a deck of cards and observe the top card. What is the probability that 12 hearts, 12 clubs, 13 diamonds, and 13 spades are observed? Find the probability that exactly 12 hearts and 12 clubs are observed. Find the probability that 25 black cards and 25 red cards are observed. (Do not evaluate.)

*5. A laboratory technician has 16 hamsters with colds. He wishes to test the effectiveness of the brands W, X, Y, and Z of cold remedies. He therefore separates them, at random, into 4 groups of 4 each, assigning the groups to the different remedies. It turns out, unknown to the technician, that 4 of the hamsters have colds that can be cured by any of the brands, while 12 have incurable colds. (The brands are equally effective.)
    a. What is the probability that the laboratory technician will discover that brand X cures all its hamsters, causing the laboratory to issue an erroneous report concerning the effectiveness of brand X?
    b. What is the probability that brand X will cure 2 hamsters, while Y and Z cure 1 apiece?
    c. What is the probability that each brand cures 1 hamster?

*6. A bag of poker chips contains 7 white chips, 8 red chips, and 14 blue chips. Five chips are drawn at random. Find the probability that
    a. 1 white, 1 red, and 3 blue chips are drawn.

b. no whites are drawn.

c. 2 red and 3 blues are drawn.

*7. How many "words" may be made out of all the letters of the word "macadem"?

*8. How many 7-digit numbers are there which use all the digits 2, 2, 2, 3, 3, 5, 5?

*9. How many terms are there in the multinomial expansion (Equation 2.24)?

*10. Evaluate $\Sigma \, 6!/a!b!c!$, where the sum is taken over all nonnegative integers (a, b, c) such that $a+b+c=6$. (*Hint:* Use Equation 2.24. Take each $x_i = 1$.)

*11. A class of 30 is divided into 3 groups, A, B, and C, at random. (Thus each student in this class is assigned the grade A, B, or C at random.) Is it more likely that C has an odd or an even number of students? (*Hint:* Compute $(1+1+-1)^{30}$ using Equation 2.24.] Generalize this problem, still using 3 classes, but making the number in the class arbitrary. Generalize, making the number of classes arbitrary. (Cf. Equation 2.18.)

*12a. Find the term involving $x^2y^3z^4$ in the expansion of $(x+y+z)^9$.
  b. Find the term involving x^2y^3 in the expansion of $(1+x+y+z)^7$.

13. If there are 80 good transistors and 20 bad transistors in a box, and if 7 transistors are chosen (without replacement) at random, what is the probability that
  a. all transistors are good?
  b. 5 are good and 2 are bad?
  c. 5 or more are good?
Do not evaluate.

14. Ten cards are chosen, without replacement, from a standard deck. Find the probability that 5 are red and 5 are black.

*6 CARDS

In this section we shall consider poker and bridge. It is necessary to start with a warning. As with most interesting games, these games are more than probability experiments. They are games of skill involving intricate rules. It is not enough to know the probabilities of various events. It is also required to know how to play. But these probabilities provide a starting point for gaining some insight into these games.

37 Poker

The first step in many poker games is to be dealt 5 concealed cards. For the purposes of poker, an ace is regarded as a 1 or 14, and a jack, queen, and king are regarded as an 11, 12, and 13, respectively. The hands are classified according to the following scheme (listed in order of decreasing value):

a. *Royal flush:* 10, J, Q, K, A of the same suit.

b. *Straight flush:* 5 consecutive ranks in the same suit, except case a, which is put into a special category.

c. *Four of a kind:* 4 cards of the same rank.

d. *Full house:* 3 cards of one rank, 2 of another.

e. *Flush:* 5 cards in the same suit, with the exception of case a or b.

f. *Straight:* 5 consecutive ranks, not all of the same suit.

g. *Three of a kind:* 3 cards of the same rank, the other 2 ranks different from each other and different from this rank.

h. *Two pair:* 2 cards of the same rank, 2 cards of a different rank, and 1 card of a rank different from these ranks.

i. *One pair:* 2 cards of the same rank, the other 3 ranks different from each other and this rank.

j. *A bust:* none of the above.

To compute the probabilities, we take as a (uniform) sample space all *unordered* samples of size 5 (*without replacement*) from the deck of 52 cards (the population). There are $\binom{52}{5} = 2{,}598{,}960$ possible hands. By computing the number of hands of type a, b, . . . , h, we may find the probability of each such hand.

For example, how many hands of type h (2 pair) are there? We may specify this hand using the following stages (framed as questions): 1. Which 2 ranks constitute the pair? 2. What are the suits of the lower pair? 3. What are the suits of the higher pair? 4. What is the odd card? Thus the hand {3 He, 3 Cl, 5 Di, 5 Cl, J Sp} is specified by the stage as follows: (1) {3, 5}, (2) {He, Cl}, (3) {Di, Cl}, (4) J Sp. The computation is as follows:

$$\underbrace{\binom{13}{2}}_{1} \times \underbrace{\binom{4}{2}}_{2} \times \underbrace{\binom{4}{2}}_{3} \times \underbrace{52-8}_{4}$$

Stage:

$$= \frac{13 \times 12}{2} \times \frac{4 \times 3}{2} \times \frac{4 \times 3}{2} \times 44 = 123{,}552$$

Thus there are 123,552 possible "2-pair" poker hands. The probability is

$$p(2 \text{ pair}) = \frac{123{,}552}{2{,}598{,}960} = .048 = 4.8\%$$

Thus one picks up 2 pair roughly 1 time in 20.

The computation of the other probabilities follows a similar pattern and is left to the exercises. Note that because *unordered* samples were considered, it was unnecessary to specify the order in which the cards were drawn. (Compare with Exercise 9, Section 2 of Chapter 2.) The probabilities satisfy a property we all suspected to be true. The higher the value of a hand, the less likely it is!

Computations such as these give us indirect experimental evidence that we are doing more than playing with numbers when probabilities are computed. They can be experimentally verified. Each student in the author's class of 31 was asked to deal himself a poker hand 5 times (shuffling well at each stage). The resulting figures are given in Table 2.17, along with the

2.17 Poker Experiment, with Probabilities

	Frequency	Relative frequency (percent)	Probability (percent)
A bust	77	46.7	50.1
One pair	73	44.2	42.3
Two pair	7	4.2	4.8
Three of a kind	4	2.4	2.1
Straight	1	.6	.4
Flush	1	.6	.2
Full house	2	1.2	.1
Other	0	.0	.0
Total	$\overline{165}$	$\overline{99.9}$	$\overline{100.0}$

theoretical probabilities (to the nearest tenth of 1 percent).

We note that in a game with 4 players, the possible initial *deals* are much larger than $\binom{52}{5}$. In fact, we have a partition of the population of 52 cards into subpopulations of size 5, 5, 5, 5, and 32 (the remaining cards). There are

$$\frac{52!}{(5!)^4 32!} = 1.48 \times 10^{24}$$

possible initial deals in a 4-handed poker game.

38 Bridge

Bridge is played with 4 people, North (N), East (E), South (S), and West (W). Each person is dealt 13 cards at random, and the game begins. We concern ourselves with initial probabilities and leave strategy to the experts.

A bridge deal is a division of a population of 52 cards into 4 subpopulations, each of size 13. Thus there are $52!/(13!)^4 = 5.36 \times 10^{28}$ possible deals. Occasionally, one reads that 4 people were playing bridge and that each player was dealt 13 cards of the same suit (a so-called perfect hand). What is the probability? The suits may be distributed in $4! = 24$ different ways. Hence the probability of such a distribution is $24/(5.37 \times 10^{28}) = 4.47 \times 10^{-28}$. This is an unlikely event.

On the other hand, what is the probability that South picks up a perfect hand? The simplest sample space is the $\binom{52}{13} = 6.35 \times 10^{11}$ (unordered) hands consisting of 13 cards. Of these, there are 4 perfect hands (1 for each suit). The probability is $4/\binom{52}{13} = 6.3 \times 10^{-12}$. This is also unlikely but about 10^{16} times more likely than the event of all players receiving such a hand.

Getting back to reality, what is the probability that a person picks up a 4-3-3-3[2] distribution? To count the possibilities, there are 4 choices for the suit with 4 cards, and, given this suit, there are $\binom{13}{4}$ ways of choosing the cards in that suit. Similarly, there are $\binom{13}{3}$ ways of choosing the cards in the other 3 suits. Thus there are $(4)\binom{13}{4}\binom{13}{3}^3$ ways of forming a 4-4-4-3 distribution. Finally, the probability of this (so-called "square") distribution is

$$p(4\text{-}3\text{-}3\text{-}3) = \frac{(4)\binom{13}{4}\binom{13}{3}^3}{\binom{52}{13}} \quad (= .1054)$$

On the other hand, similar reasoning gives

$$p(5\text{-}3\text{-}3\text{-}2) = \frac{(4 \cdot 3)\binom{13}{5}\binom{13}{3}^2\binom{13}{2}}{\binom{52}{13}}$$

Dividing these 2 probabilities, we have, after simplification,

$$\frac{p(5\text{-}3\text{-}3\text{-}2)}{p(4\text{-}3\text{-}3\text{-}3)} = \frac{81}{55} = 1.47$$

2 The numbers refer to the number in each suit. However, the order of the suits are not specified. The 4 may refer to any of the suits.

Thus a 5-3-3-2 distribution is about $1\frac{1}{2}$ times as likely as the 4-3-3-3 distribution.

*EXERCISES

In the following exercises, evaluate the answers only where feasible — i.e., where the numbers are relatively small, or where cancellations simplify a problem, or where curiosity and (if necessary) a table of logarithms are at hand.

1. Find the number of possibilities for each of the poker hands a to j of the text, and compute the probabilities.

2. A poker player picks up a hand with 2 kings and 3 other cards (not kings and with different ranks). The rules permit him to exchange these cards for 3 others in the deck. He makes this exchange. What is the probability that he improves his hand?

3. As in Exercise 2, a poker player picks up 3 spades (3, 7, J) and 2 cards of another suit. He exchanges these 2 cards for 2 other cards in the deck. What is the probability that he draws a flush in spades? What is the probability that he gets 1 pair or better? (You may assume that he did not exchange a 3, 7, or J.)

4. Some people play poker with deuces wild. The means that a deuce (2) may be regarded, at the player's option, as any card (rank and suit arbitrary) that the player specifies. In this case 5 of a kind is a possible poker hand. In a deuces-wild game, which is more likely — a royal flush or 5 of a kind?

5. "Poker" may be played with 5 dice. Compute the probability of the analogues of
  a. 5 of a kind.   **b.** 4 of a kind.
  c. a full house.   **d.** a straight.
  e. 3 of a kind.   **f.** 2 pair.
  g. 1 pair.   **h.** a bust.
Arrange in order of probability — the least likely first.

6. In bridge, find the probability that a player does not pick up an ace.

7. In bridge, find the probability that all 4 players have aces.

8. In bridge, find the probability that a player picks up 11 or more cards in the same suit.

9. In 7-card stud a player is dealt 7 cards and he chooses the 5 cards that give him the best hand. What is the probability of a royal flush? A straight flush? A flush? A full house? Two pair?

CHAPTER 3 GENERAL THEORY OF FINITE PROBABILITY SPACES

INTRODUCTION

If S is any probability space (Definition 1.6), we have defined an event A as any subset of S (Definition 1.12). In Chapter 2 we worked with a uniform space S and considered the problem of computing the probability $p(A)$ by counting the elements in A and in S and using Equation 1.11: $p(A) = n(A)/n(S)$.

In this chapter we shall learn some of the techniques for computing probabilities of events in an *arbitrary* finite probability space. It is well to keep in mind the purpose of many of the theorems we study. Simply put, we compute probabilities of complicated events by using the probabilities of the simpler events as a basis. Similarly, we build complicated sample spaces (these are the ones in real life) out of building blocks consisting of simple sample spaces. This chapter may well be subtitled "New Probabilities from Old."

1 UNIONS, INTERSECTIONS, AND COMPLEMENTS

In what follows, we shall assume that we have a fixed sample space $S = \{s_1, \ldots, s_k\}$, with $p(s_i) = p_i$. *All events, or sets, are understood to be subsets of S.* Much of what is stated applies to infinite spaces as well.

If A and B are sets, there are two fundamental operations that we can perform on A and B and one operation on A to form a new set.

1 Definition

If A and B are sets, the *union* of A and B, denoted $A \cup B$, is the set of points belonging to A or to B or to both A and B. The *intersection* of A and B, denoted $A \cap B$, is the set of points belonging to both A and to B. The *complement* of A, denoted $\bar{A}$, is the set of points not belonging to A (but belonging to S).

Figure 3.1 illustrates these definitions.

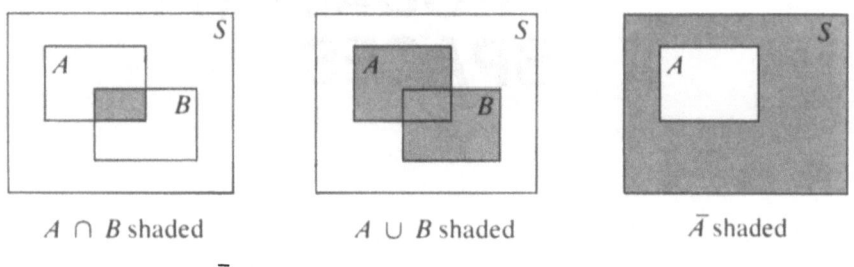

$A \cap B$ shaded $A \cup B$ shaded $\bar{A}$ shaded

3.1 $A \cap B, A \cup B, \bar{A}$

2 Example

In the game of black jack the dealer is dealt 2 cards, 1 concealed and 1 exposed. Let S be the set of all possible hands. Thus S is the set of ordered samples (size 2) from the population of 52 cards. Suppose we define A to be the event "The hidden card is an ace." This means, in sample-space terminology, that A is the set of all hands in which the hidden card is an ace. Suppose B is the event "The exposed card is a 10, J, Q, or K." Then we may define several events in terms of A and B. For example, $A \cap B$ is the event "The hidden card is an ace, and the exposed card is a 10, J, Q, or K." $\bar{A} \cap B$ is "The hidden card is not an ace, but the exposed card is a 10, J, Q, or K." $A \cup B$ is "The hidden card is an ace, or the exposed card is a 10, J, Q, or K." $\bar{B}$ is "The exposed card is not a 10, J, Q, or K."

Some events cannot be expressed in terms of A and B by using intersections, unions, or complements. Thus "The exposed card is an ace" is such an event.

A simple rule of thumb determines whether unions, intersections, or complements are involved:

The word "and" connecting two events usually suggests an intersection.

The word "or" suggests a union. (We always understand the word "or" to be the same as the term "and/or." Thus when we say that the car is red or the chair is blue, we include the case that both the car is red and the chair is blue. This is the universal meaning of "or" in mathematics, and it is the usual meaning in the English language.)

The word "not" suggests a complement.

For this reason a perfectly plausible terminology is to use "A and B" for "$A \cap B$," "A or B" for "$A \cup B$," and "not A" for "$\bar{A}$." However, the notation we use is now rather standard.

The following results are useful for the computation of the probabilities of events that are built up from other more simpler events.

3 Definition

The events A and B are said to be *mutually exclusive* if $A \cap B = \emptyset$.

Thus A and B are mutually exclusive if they cannot occur simultaneously. For example, if S is a pack of cards, $A = \{$red cards$\}$ and $B = \{$spade picture cards$\}$, then A and B are mutually exclusive.

4 Theorem. On Mutually Exclusive Events

$$p(A \cup B) = p(A) + p(B) \qquad \text{if } A \text{ and } B \text{ are mutually exclusive} \qquad (3.1)$$

5 Theorem. On Complements

$$p(\bar{A}) = 1 - p(A) \qquad (3.2)$$

To prove Theorem 4 we go to the definition of the probability of an event (Definition 12 of Chapter 1). Writing $A = \{a_1, \ldots, a_s\}$, $B = \{b_1, \ldots, b_t\}$, we have $A \cup B = \{a_1, \ldots, a_s, b_1, \ldots, b_t\}$. In the latter set the a_i's and b_i's are all distinct, because A and B are mutually exclusive. By Definition 12 of Chapter 1 we have

$$p(A \cup B) = p(a_1) + \cdots + p(a_s) + p(b_1) + \cdots + p(b_t)$$
$$= p(A) + p(B)$$

This proves Theorem 4.

To prove Theorem 5 we note that A and $\bar{A}$ are mutually exclusive and that $A \cup \bar{A} = S$. (This is an immediate consequence of the definition of $\bar{A}$.) Thus $p(S) = p(A) + p(\bar{A})$ by Theorem 4. But, by Equation 1.13, $p(S) = 1$. Thus we obtain

$$p(A) + p(\bar{A}) = 1 \qquad (3.3)$$

which is equivalent to Equation 3.2.

Equation 3.2 may be generalized. We say that $A \subseteq B$ if all elements of A are also in B. In this case we let $B-A$ designate the set of elements that are in B but not in A. $(B-A = B \cap \bar{A}.)$ Then we have

$$p(B-A) = p(B) - p(A) \qquad (A \subseteq B) \qquad (3.4)$$

To see this, we note that $B = A \cup (B-A)$ and that A and $B-A$ are mutually exclusive. (Thus B is the union of A and the set of points of B that are not in A.) Hence $p(B) = p((A) \cup (B-A)) = p(A) + p(B-A)$ by Equation 3.1. Transposing $p(A)$ we obtain Equation 3.4. Equation 3.2 is the special case $B = S$.

Finally, we may generalize Theorem 4 to arbitrary sets A and B.

6 Theorem

For any events A and B,

$$p(A \cup B) = p(A) + p(B) - p(A \cap B) \qquad (3.5)$$

To prove this we note that $A \cap B \subseteq B$. Furthermore, $A \cup B = A \cup (B - A \cap B)$. (Thus, to form $A \cup B$ we adjoin to A the points of B that are not already in A.) Hence we have, by Equations 3.1 and 3.4,

$$\begin{aligned} p(A \cup B) &= p(A \cup (B - A \cap B)) \\ &= p(A) + p(B - A \cap B) \\ &= p(A) + p(B) - p(A \cap B) \end{aligned}$$

We may regard Equation 3.5 as follows. To form the probability of $A \cup B$ we add the probability of A to the probability of B. But certain sample points — those which belong to A and B — are counted twice. Hence it is necessary to subtract $p(A \cap B)$.

7 Example

In a certain class, 35 percent of the students are hard working, 50 percent are smart, and 10 percent are hard working and smart. What percentage is either hard working or smart? What percentage is neither hard working nor smart?

If H is the set of hard-working students and S is the set of smart students, we have $p(H) = .35$, $p(S) = .50$, $p(H \cap S) = .10$. Thus $p(H \cup S) = p(H) + p(S) - p(H \cap S) = .50 + .35 - .10 = .75$. Thus 75 percent are in the smart or hard-working category. The complement of this group is the group of students that are neither smart nor hard working. Thus

$$p(\bar{H} \cap \bar{S}) = p(\overline{H \cup S}) = 1 - p(H \cup S) = 1 - .75 = .25$$

This example illustrates the following general properties of sets, known as *De Morgan's laws*:

$$\overline{A \cup B} = \bar{A} \cap \bar{B} \qquad \overline{A \cap B} = \bar{A} \cup \bar{B} \qquad (3.6)$$

By drawing a diagram similar to Fig. 1.1, the reader may readily convince himself of the truth of Equations 3.6.

Example 7 is not a traditional "probability problem." We are using the idea, given in Section 5 of Chapter 1, that a finite set may be regarded as a uniform sample space.

It is possible to generalize the above considerations to more than two sets.

8 Definition

If $A_1, \ldots, A_s$ are sets, the *union* $A_1 \cup \cdots \cup A_s$ is defined as the set of points belonging to at least one A_i. The *intersection* $A_1 \cap \cdots \cap A_s$ is the set of points belonging to all the A_i's.

Definition 1 is the special case $s = 2$. We may now easily generalize Theorem 4.

9 Definition

The sets $A_1, \ldots, A_s$ are *mutually exclusive* if any 2 of these sets are mutually exclusive: $A_i \cap A_j = \emptyset$ for $i \neq j$.

10 Theorem. On Mutually Exclusive Events

$$p(A_1 \cup \cdots \cup A_s) = p(A_1) + \cdots + p(A_s)$$
$$\text{if the sets } A_i \text{ are mutually exclusive} \qquad (3.7)$$

The proof is a straightforward extension of the proof of Theorem 4, and we omit it.

11 Example

Twenty coins are tossed. What is the probability that 17 or fewer heads turn up?

As in Example 2.25, the probability that exactly r heads turn up is $\binom{20}{r}/2^{20}$. If E_r is the event that exactly r heads occur, we are looking for the probability of the event

$$E = E_0 \cup E_1 \cup \cdots \cup E_{17}$$

However, the numbers are smaller and fewer if $\bar{E}$ is considered. $\bar{E}$ is the event that 18 or more heads occur. Since the sets E_r are clearly mutually exclusive, we may use Equation 3.7 to obtain

$$p(\bar{E}) = p(E_{18}) + p(E_{19}) + p(E_{20})$$
$$= \left[\binom{20}{18} + \binom{20}{19} + \binom{20}{20} \right]/2^{20}$$
$$= (190 + 20 + 1)/2^{20}$$
$$= 211/2^{20} \ (= .00020)$$

Finally, $p(E) = 1 - p(\bar{E}) = .99980$ (to 5 decimal places).

Equations 3.1 through 3.5 have their counterparts in counting finite sets. Using Equations 1.14 and 1.15 we can convert any statement about probabilities into one about numbers. All that is required is the number of elements in the set, which we regard as a uniform sample space. Thus we have the following theorem.

12 Theorem

Let S be a set with k elements, and let A and B be subsets of S. Then

$$n(A \cup B) = n(A) + n(B) \qquad \text{if } A \text{ and } B \text{ are mutually exclusive} \qquad (3.8)$$

$$n(\bar{A}) = k - n(A) \qquad (3.9)$$

$$n(B - A) = n(B) - n(A) \qquad \text{if } A \subseteq B \qquad (3.10)$$

and

$$n(A \cup B) = n(A) + n(B) - n(A \cap B) \qquad (3.11)$$

For a proof, it is sufficient to multiply Equations 3.1, 3.2, 3.4, and 3.5 by k, and to use Equation 1.15.

Equation 3.8 is so basic that it is often used to define the sum of 2 numbers in a theoretical formulation of arithmetic. We used it in Example 1.14, where A and B were called the "cases." Equation 3.11, a generalization, even permits the "cases" to overlap.

13 Example

A player draws 3 cards (without replacement) from a deck. He wins if they are all the same color or if they form a straight (3 consecutive ranks). What is his probability of winning?

We use the uniform probability space of unordered triples. There are $k = \binom{52}{3} = 52 \cdot 51 \cdot 50/3 \cdot 2 \cdot 1 = 22{,}100$ elementary events. We let C be the set of hands with cards of the same color and S the set of hands that form straights. (As usual, we interpret an ace as a 1 or a 14.) If W is the set of winning hands, $W = S \cup C$ and

$$n(W) = n(S) + n(C) - n(S \cap C) \qquad (3.12)$$

A straight is determined by the starting point and then the suits for the 3 cards. Since we may start a straight anywhere from A through Q (1 through 12), we have $n(S) = 12 \times 4^3 = 768$. A hand in C is specified by the color (red or black) and then a selection of 3 cards of that color. Hence $n(C) = 2 \times \binom{26}{3} = 5{,}200$. Finally, a hand of $S \cap C$ is specified by (1) the color, (2) the beginning rank, and (3) the 3 suits (in the color selected) from high to low. Hence $n(S \cap C) = 2 \times 12 \times 2^3 = 192$. Thus, by Equation 3.12, we have

$$n(W) = 768 + 5{,}200 - 192 = 5{,}776$$

and finally

$$p(W) = \frac{n(W)}{n(S)} = \frac{5,776}{22,100} \quad (= .261)$$

EXERCISES

1. Let S be the integers from 1 through 10 inclusive. Let A be the odd integer of S, let B be the primes $\{2, 3, 5, 7\}$ of S, and let C be the numbers from 3 through 8. Find

a. $A \cup B$ b. $\bar{B}$

c. $A \cap \bar{C}$ d. $A \cap B \cap C$

e. $A \cap \bar{B} \cap \bar{C}$ f. $\bar{A} \cap B \cap C$

g. $A \cup B \cup C$ h. $\overline{A \cup B \cup C}$

2. In Exercise 1 verify Equation 3.11 for the sets A and B, for B and $\bar{C}$, and for A and C.

3. Let S be the set of integers (positive, negative, or zero), P the set of positive integers, E the set of even integers, T the multiples of 3, and Sq the set $\{0, 1, 4, 9, \ldots\}$ of squares. Express each of the following, if possible, using unions, intersections, or complements. (*Warning:* Some may not be possible.)

a. The odd integers.   b. The positive even integer.

c. The odd squares.   d. All multiples of six.

e. The nonnegative integers.   f. All squares larger than 2.

g. $\{0\}$.   h. $\{0, 1\}$.

4. Suppose $A \triangle B$ is defined to be the set of elements in A or in B but *not* in both. Derive a formula for $p(A \triangle B)$.

5. Let S be the sample space given in the following table:

s	a	b	c	d	e	f	g
p	.10	.20	.05	.20	.30	.05	.10

Let $A = \{a, b, c\}$, $B = \{a, c, e, f\}$, $C = \{c, d, e, f, g\}$.

a. Find $\bar{A}, \bar{B}, \bar{C}, A \cup B, B \cap C, A \cap B \cap C, \bar{A} \cup B$.

b. Find the probabilities of the events in part a.

c. Verify Theorem 6 for $p(A \cup C)$.

d. Verify Equation 3.6 for $\overline{B \cup C}$ and for $\overline{A \cap B}$.

6. There are 20 apples in a bag which are either red or rotten. There are

12 red apples and 13 rotten ones. How many red, rotten apples are there? State specifically which formula you are using, and identify your sets.

7. It is observed that 7 percent of the words on a certain page begin with the letter "e" and 6 percent end with "e." Suppose that 1 percent begin and end with "e." What percentage begins or ends with "e"?

8. Two investigators are given a very difficult code to crack. It is known from past experience that each investigator has probability .18 of cracking such a code. What can you say about the probability that the code will be cracked? Explain.

9. Three dice are tossed. What is the probability that at least one 6 occurs? Similarly, find the probability of at least one 6 when 4 dice are tossed. (*Hint:* Look at the complement.)

10. Three dice are tossed. What is the probability that some number will be repeated? Do the same for 4 dice.

11. An integer from 1 to 100 inclusive is selected at random. What is the probability that it is either a square or a multiple of 3?

12. Ten coins are tossed. What is the probability that the number of heads is between 3 and 7 inclusive? (You may use Table 2.13.)

13. How many integers between 1 and 600 inclusive are divisible by either 3 or 5?

14. An urn contains 7 black, 8 red, and 9 green marbles. Five marbles are selected (without replacement) at random. What is the probability that some color will be missing? (Do not evaluate.)

15a. Two cards are chosen from a deck. What is the probability that at least 1 ace is chosen or that both have the same color?
b. If 3 cards are chosen, what is the probability that an ace is chosen or that the cards have the same color?

16. A packet of seeds contains 100 seeds, of which 85 will germinate. Ten seeds are chosen at random and planted. What is the probability that 7 or more will germinate? (Do not evaluate.)

17. If $p(A)$ and $p(B)$ are known, then the value of $p(A \cup B)$ cannot be exactly determined, because Equation 3.5 involves the unknown value of $p(A \cap B)$. In each of the following, determine how large and how small $p(A \cup B)$ can be.

$\quad$ **a.** $p(A) = .1, p(B) = .2$ $\qquad$ **b.** $p(A) = .5, p(B) = .4$
$\quad$ **c.** $p(A) = .6, p(B) = .7$ $\qquad$ **d.** $p(A) = .2, p(B) = .9$
$\quad$ **e.** $p(A) = .9, p(B) = .95$

3.2 Unemployed Persons, Age 20 Years and Over, by Age and Sex, 1960

	20–24 yr		25–44 yr		45–64 yr		65 and over		Total	
	Male	Female	Male	Female	Male	Female	Male	Female	Male	Female
Number (in thousands)	369	214	907	516	686	323	96	25	2,058	1,078
Percentage	11.8	6.8	28.9	16.5	21.9	10.3	3.1	0.8	65.6	34.4

18. Give a rule that determines how large and how small $p(A \cup B)$ can be in terms of the values of $p(A)$ and $p(B)$.

19. In Exercise 17 determine how large and how small $p(A \cap B)$ can be.

20. In analogy with Exercise 18, estimate the size of $p(A \cap B)$ in terms of the size of $p(A)$ and $p(B)$.

2 CONDITIONAL PROBABILITY

Table 3.2 is adopted from government sources[1] and gives data on unemployment by sex and age. We may regard Table 3.2 as giving the results of an "experiment" that had 8 possible outcomes.

If we wish to analyze the unemployed *males* by age, we may use the appropriate entries in Table 3.2 to construct a new table. In this case we *restrict* the sample space to the event "male" and we arrive at Table 3.3.

3.3 *Unemployed Males, Age 20 Years and Over, by Age and Sex, 1960*

	20–24 yr	25–44 yr	45–64 yr	65 and over	Total
Number (in thousands)	369	907	686	96	2,058
Percentage	17.9	44.1	33.3	4.7	100.0

When the outcomes of an experiment are restricted in this manner to a specified event, we speak of the *conditional relative frequency* of an outcome or an event. We have used the notation $f(A)$ or $f(s)$ for the relative frequency of an event or outcome. We shall use the notation $f(A|B)$ (*read: f of A, given B*) to designate the relative frequency of the event A when the event B is regarded as the total sample space. For example, to compute $f(25–44|$ male$)$ we read, from Table 3.2 or 3.3, that 907,000 males are in the age bracket 25–44, and that there are 2,058,000 males. Thus $f(25–44|$male$) = 907/2,058 = .441$, as indicated in Table 3.3. Note that the numerator 907 was *not* the total in the 25–44 category — it was the total in the (25–44) $\cap$ male category! We now generalize this procedure.

1 U.S. Bureau of the Census, *Statistical Abstract of the United States: 1962* (eighty-third edition), Government Printing Office, Washington, D.C., 1962.

14 Definition

Suppose an experiment has outcomes $s_1, \ldots, s_k$ and that A and B are events. The *relative frequency* of A, given B, is defined by the equation

$$f(A|B) = \frac{n(A \cap B)}{n(B)} \qquad (3.13)$$

Here we again use the notation $n(B)$ to denote the number of occurrences (the frequency) of B. Again, the numerator is not $n(A)$, but $n(A \cap B)$. Equation 3.13 is not defined if $n(B) = 0$. Nor do we wish to define the relative frequency of an event A given an event B that has not occurred.

To illustrate Definition 14 again, with a different base population B, let us find the percentage of males among the unemployed in the age category 20–44. Referring to Table 3.2, we find $n(20\text{–}44) = 369 + 214 + 907 + 516 = 2{,}006$, while $n(\text{males} \cap (20\text{–}44)) = 369 + 907 = 1{,}276$. Thus $f(\text{males}|(20\text{–}44)) = n(\text{males} \cap (20\text{–}44))/n(20\text{–}44) = 1{,}276/2{,}006 = 63.6$ percent. Among the population of unemployed in the age group 20–44, 63.6 percent are males.

Just as the relative frequency of an event may be computed from the relative frequencies of the outcomes in it (Theorem 1.10), we can compute the conditional relative frequency $f(A|B)$ directly from relative frequencies.

15 Theorem

Let A and B be events in some experiment with the relative frequency $f(B) \neq 0$. Then

$$f(A|B) = \frac{f(A \cap B)}{f(B)} \qquad (3.14)$$

Proof. If a total of N experiments are performed, then, by Definition 1.9,

$$f(B) = \frac{n(B)}{N} \qquad \text{and} \qquad f(A \cap B) = \frac{n(A \cap B)}{N}$$

Dividing these equations, we obtain, using Equation 3.13,

$$\frac{f(A \cap B)}{f(B)} = \frac{n(A \cap B)/N}{n(B)/N} = \frac{n(A \cap B)}{n(B)} = f(A|B)$$

This is the result.

To illustrate this theorem, let us compute $f(\text{males}|(20\text{–}44))$ once again. We have, using Table 3.2, $f(20\text{–}44) = .118 + .068 + .289 + .165 = .640$, $f(\text{males} \cap (20\text{–}44)) = .118 + .289 = .407$. Hence, by Equation 3.14, $f(\text{males}|(20\text{–}44)) = .407/.640 = 63.6$ percent, as before.

Since probabilities may be closely approximated by relative frequencies if a large number of trials are performed, Theorem 15 suggests that the *probability* of A, given B, be defined.

16 Definition

Let S be a probability space, and let A and B be events of S with $p(B) \neq 0$. Then the *probability* of A, given B, or the conditional probability of A, given B, is defined by the formula

$$p(A|B) = \frac{p(A \cap B)}{p(B)} \tag{3.15}$$

In the light of Equation 3.14, we see that the conditional probability $p(A|B)$ is approximated by the conditional relative frequency $f(A|B)$ if a large number of trials are performed. We might also note, as indicated in Section 5 of Chapter 1, that any table of frequencies may be viewed as a probability space. From this point of view, Equation 3.14 is a special case of Equation 3.15.

We have pointed out that behind any probability there is, in theory, an experiment. It is well to state the experiment corresponding to $p(A|B)$.

If A and B are events in an experiment with sample space S, and if $p(B) \neq 0$, then to find an experimental value for $p(A|B)$, we repeat the experiment many times, ignoring all outcomes except those in B, and find the relative frequency of the outcomes in B that are also in A.

17 Example

A die is tossed repeatedly. What is the (conditional) probability that 6 turns up on or before the fourth toss, given that it turns up somewhere in the first 10 tosses?

We refer to Table 1.6 and take for granted the probabilities in that table. Let A be the event "4 or less throws required" and let B be the event "10 or less throws required." We wish to find $p(A|B)$. Here $A \cap B = A$, because any occurrence of A implies an occurrence of B. Thus $p(A|B) = p(A)/p(B)$. From Table 1.6 we read $p(A) = .167 + .139 + .116 + .096 = .518$ and $p(B) = 1 - (.027 + .022 + .112) = 1 - .161 = .839$. [Here it is easier to compute $p(\bar{B})$ and then $p(B)$.] Thus $p(A|B) = .518/.839 = .617$.

The (unconditional or absolute) probability of tossing a 6 in 4 or less trials is $p(A) = .518$. We see that, in this case, it was intuitively clear that $p(A|B)$ was larger than $p(A)$, because if we failed to toss a 6 in the first 4 attempts, it was still possible to avoid a 6 in the next 6 attempts, so the failure would be "washed out."

In this example $p(A|B) = p(A)/p(B)$, because A was included in B. (If A occurred, B occurred automatically.) For later reference we record this special case of Equation 3.15:

$$p(A|B) = \frac{p(A)}{p(B)} \qquad \text{if } A \subseteq B \tag{3.16}$$

If $B = S$ (the entire sample space), we automatically have $A \subseteq S$. Also,

$p(S) = 1$. Thus

$$p(A|S) = p(A) \tag{3.17}$$

Thus any probability may be regarded as a conditional probability, given S.

18 Example

Two cards are chosen from a deck. At least 1 of the cards is red. What is the probability that both are red?

We let A be the event that both are red, and we let B be the event that at least 1 is red. Again $A \subseteq B$. We use ordered samples to compute the probabilities. $p(A) = 26 \cdot 25/52 \cdot 51 = \frac{25}{102}$. $\bar{B}$ is the event that both are black. Thus $p(\bar{B}) = \frac{25}{102}$, as with $p(A)$, and hence $p(B) = 1 - \frac{25}{102} = \frac{77}{102}$. Finally, using Equation 3.16, we have $p(A|B) = \frac{25}{102}/\frac{77}{102} = \frac{25}{77} = .325$.

Another way of obtaining this answer is as follows. There are $52 \cdot 51$ possible hands. Of these, $26 \cdot 25$ hands are all black. Hence there are $52 \cdot 51 - 26 \cdot 25 = n(B)$ hands with at least 1 red card, and all of these are equally likely. Of these hands, there are $26 \cdot 25 = n(A)$ all-red hands. Thus

$$p(A|B) = \frac{26 \cdot 25}{52 \cdot 51 - 26 \cdot 25} = \frac{25}{102 - 25} = \frac{25}{77}$$

This technique may be used for uniform probability spaces.

19 Theorem

If S is a finite *uniform* probability space and if $B \neq \emptyset$, then

$$p(A|B) = \frac{n(A \cap B)}{n(B)} \tag{3.18}$$

The proof follows directly from Equation 3.15, using Formula 1.11, which is valid for uniform spaces. The details are left to the reader.

20 Example

Two cards are chosen from a deck. The first is red. What is the probability that the second is also red?

Since the problem is phrased in terms of first and second cards, ordered samples are called for. We let R_1 be the event "first card is red" and we let R_2 be the event "second card is red." Then it is required to find

$$p(R_2|R_1) = \frac{p(R_2 \cap R_1)}{p(R_1)}$$

We have $p(R_2 \cap R_1) = 26 \cdot 25/52 \cdot 51$, while $p(R_1) = 26 \cdot 51/52 \cdot 51$. Thus

$$p(R_2|R_1) = \frac{26 \cdot 25}{52 \cdot 51} \bigg/ \frac{26 \cdot 51}{52 \cdot 51} = \frac{26 \cdot 25}{52 \cdot 51} \cdot \frac{52 \cdot 51}{26 \cdot 51} = \frac{25}{51}$$

Several interesting questions arise in connection with this problem. First, the answer $\frac{25}{51}$ is obvious. Clearly if the first card was red, the probability that the second card is red is $\frac{25}{51}$, because there are 25 red cards among the 51 (equally likely) cards left over, regardless of what the first red card is. We shall justify this technique in Section 3.5, but for the present we note that the obvious answer $\frac{25}{51}$ was verified in Example 20. In what follows we shall use the "obvious" answer without further justification.

Similarly, another "obvious" computation above was $p(R_1) = \frac{1}{2}$. Indeed, of the 52 (equally likely) cards in the deck, there were 26 red cards. Yet our computation used ordered couples: There were 26 red cards (first stage), then 51 cards (second stage), etc. Why do it this way? We have stressed the importance of the *probability space* throughout this text, and it is crucial that we understand that *all* events are taken to be subsets of a given probability space. The underlying reason that the simple computation $[p(R_1) = \frac{26}{52}]$ works is our choice of a uniform probability space for the possible hands. We shall also clarify this in Section 5, but we shall henceforth use the simpler method, when applicable.

Many people who write $p(R_1) = \frac{1}{2}$ without thought balk at the equation $p(R_2) = \frac{1}{2}$. The common attitude is that $p(R_2)$ "depends on what color the first card is." However, this is needless balking, because the same line of reasoning would show that $p(R_1)$ depends similarly on what color the second card is! The situations are, in fact, identical except for a time sequence. Even this time sequence is an illusion, for it is somewhat arbitrary which card is called the first. In fact, $p(R_1)$ and $p(R_2)$ do not "depend on the other card." They are not conditional probabilities.

Our final example is also taken from cards. The results are somewhat surprising at first glance.

21 Example

Two cards are chosen at random from a deck. One is a king. What is the probability that both are kings? If one is a king of spades, what is the probability that both are kings?

To answer the first part, there are $\binom{52}{2} = 1{,}326$ possible hands. Of these there are $\binom{48}{2} = 1{,}128$ hands without kings. Thus there are $1{,}326 - 1{,}128 = 198$ hands with at least 1 king present. Clearly, there are $\binom{4}{2} = 6$ hands with 2 kings. Thus, by Theorem 19, the required conditional probability is

$$p(\text{both kings} \mid \text{one is a king}) = \tfrac{6}{198} \ (= 3.03\%)$$

The calculation for the second part is similar. There are 51 hands with the king of spades. Of these, 3 have 2 kings. Thus the required probability here is

$$p(\text{both kings}|\text{one is the king of spades}) = \tfrac{3}{51} \ (= 5.88\%)$$

almost double the first answer.

We have noted in Example 20 that the computation $p(\text{both cards red}|\text{first card is red})$ can be computed by taking a specific instance (say the 3 of hearts) for the first card. However, $p(\text{both cards kings}|\text{one is a king})$ *cannot* be computed by taking a specific instance (say the king of spades). We shall go into this in detail in Section 5, but for the present we may say that the reason for this disparity is that in Example 20 the specific instances all yielded the same answer *and were mutually exclusive*. In Example 21, the four specific instances (one is the king of spades, one is the king of clubs, etc.) all yielded the same answer but were *not* mutually exclusive. It is this feature that distinguishes the two problems.

In this section we defined the conditional probability $p(A|B)$ of an event A relative to an event B. We have regarded it, intuitively, as a certain kind of probability of the event A. Yet we have not discussed the underlying probability space and the probabilities of its elementary events in such a way as to make $p(A|B)$ a bona fide probability of an event, in the sense of Definition 12 of Chapter 1. This will be done in Section 3.5, and the interested reader can refer to Definition 39 of that section to see how this is done.

EXERCISES

1. In Table 3.2 find the percentage of females in the 45-and-over age category. Use both Definition 14 and Theorem 15.

2. When 3 dice are tossed, what is the probability that the high number is 3 or under, given that it is 5 or under? (Use the results of Table 1.3.)

3. Using Table 1.12, find the probability that the sum 7 occurs when 2 dice are tossed, given that a sum between 5 and 9 inclusive occurred.

4. Using Table 2.17, find the probability of receiving "2 pair" in poker, given that a "bust" did not occur. Compare with the conditional relative frequency using the experimental results of that table.

5. Show that if $A \subseteq B$ and $p(B) \neq 0$, the conditional probability $p(A|B)$ is greater than or equal to the "absolute" probability $p(A)$. When does equality hold?

6. Find the probability of tossing the sum 9 or more on 2 dice given that 1 of the dice is 3 or more. Find the probability, given that the first die is 3 or more.

7. In a poker hand (5 cards), what is the probability of exactly 2 aces, given that there is 1 or more ace in the hand? (Do not evaluate.)

8. An urn contains 10 black balls and 5 red balls. Three balls are chosen (without replacement) at random. Given that they are not all red, find the probability that they are all black.

9. Three cards are chosen at random from a deck. Find the probability that they are all aces. Find the probability that they are all aces given that they all have different suits. Find the probability that they are all aces, given that they are not all the same color.

10. In applying to schools A and B, Leonard decided that the probability that he will be accepted by A or B is .7 or .8, respectively. He also feels that there is a probability .6 of being accepted by both.

    a. What is his probability of being accepted by at least one of the schools?

    b. If he is accepted by B, what is the probability that he will also be accepted by A?

11. If a 5-card poker hand has cards of the same color, what is the conditional probability that they are of the same suit?

12. A soap firm knows that 23 percent of its customers are magazine readers, 72 percent are television viewers, and 12 percent are magazine readers and television viewers. If a customer is found to be a magazine reader, what is the probability that he is a television viewer? Suppose he is a television viewer. What is the probability that he reads magazines?

13. If B has a high probability of occurring, then $p(A|B)$ is almost the same as $p(A)$. Illustrate this statement for the case $p(B) = .9$, $p(A) = .6$. Find upper and lower bounds[2] for $p(A|B)$. Do the same for $p(A) = .6$ and $p(B) = .99$. (*Hint:* See Exercises 19 and 20 of the previous section.)

14. An 8-card deck is composed of the 4 aces and the 4 kings. Two cards in a specific order are chosen at random. Find the probability that both are aces, given that

    a. the first card is an ace.

    b. the hand contains the ace of spades.

    c. the hand contains an ace.

15. Describe an experiment that may be performed to find, experimentally, the probabilities described in Exercise 14b and c. Run the experiment to obtain 50 readings on the probability in Exercise 14c. Keep track of the readings relevant to Exercise 14b, and in this manner obtain experimental conditional probabilities for Experiment 14b and c.

---

2 Thus find how large $p(A \mid B)$ can possibly be and, similarly, how small it may be.

3 PRODUCT RULE

The notion of conditional probability has wide application in practical situations. For example, to determine insurance rates, insurance companies find the probability that a driver will get involved in an auto accident. Experience shows them that the probabilities vary according to age, sex, experience, etc. Thus they effectively reduce the sample space and are dealing with conditional probabilities. (The rate structure reflects this fact.) Similarly, if a school wishes to find the probability that a student receives grade B or better, a thorough analysis would find this probability for freshmen, sophomores, etc., and perhaps according to other categories. Again, conditional probabilities are involved, because the sample space is restricted. In this section we learn how to apply information about conditional probability.

22 Theorem. The Product Rule

$$p(A \cap B) = p(A) \cdot p(B|A) \tag{3.19}$$

This formula is merely another version of Equation 3.15. (The roles of A and B are reversed, and it has been cleared of fractions.) It is used to compute $p(A \cap B)$ with the help of a known conditional probability $p(B|A)$.

23 Example
In a certain high school, 40 percent of the seniors will go to college next year. The seniors are 18 percent of the school population. What percentage of the school will go to college next year?

If we let S = seniors and C = college-bound students, we have $p(C|S) = .40$ and $p(S) = .18$. Thus

$$p(S \cap C) = p(S) \cdot p(C|S) = (.18)(.40) = .072 = 7.2\%$$

(We are assuming that only the seniors can go to college next year.)

In this example we interpreted relative frequency as a probability. This is permitted, as discussed in Section 5 of Chapter 1.

24 Example
Two cards are chosen from a deck. What is the probability that the first is an ace and the second is a king?

If we let A_1 be the event "first card is an ace" and let K_2 be the event "second card is a king," we wish to find $p(A_1 \cap K_2)$. By the product rule we have

$$p(A_1 \cap K_2) = p(A_1) \cdot p(K_2|A_1) = \left(\tfrac{4}{52}\right)\left(\tfrac{4}{51}\right) = \tfrac{4}{663}$$

[We may also find $p(K_2 \cap A_1) = p(K_2) \cdot p(A_1|K_2)$ to arrive at the same answer. There is no "time sequence" implied in the product rule.] The computation is very similar to previous computations. Thus we may say that there are $4 \cdot 4$ ways of succeeding out of $52 \cdot 51$ hands. But the above procedure is more natural, because it involves simple probabilities at each stage.

25 Example

Urn I contains 5 red and 3 green balls. Urn II contains 2 red and 7 green balls. One ball is transferred (at random) from urn I to urn II. After stirring, 1 ball is chosen from urn II. What is the probability that the (final) ball is green?

This is a 2-stage process, and it is natural to proceed by cases. Let R_1 be the event that a red ball was transferred, and let G_1 be the event that a green ball was transferred. Similarly, let G_2 be the event that the final ball is green. The computations of the conditional probabilities are easy[3]: $p(G_2|R_1) = \frac{7}{10}$, $p(G_2|G_1) = \frac{8}{10}$. We also have $p(R_1) = \frac{5}{8}$, $p(G_1) = \frac{3}{8}$. Using these equations, we can find $p(R_1 \cap G_2)$ and $p(G_1 \cap G_2)$. Thus

$$p(R_1 \cap G_2) = p(R_1) \cdot p(G_2|R_1) = (\tfrac{5}{8})(\tfrac{7}{10}) = \tfrac{35}{80}$$
$$p(G_1 \cap G_2) = p(G_1) \cdot p(G_2|G_1) = (\tfrac{3}{8})(\tfrac{8}{10}) = \tfrac{24}{80}$$

Finally, the event G_2 can happen in two mutually exclusive ways: $R_1 \cap G_2$ and $G_1 \cap G_2$. Thus, by Equation 3.4, we have

$$p(G_2) = \tfrac{35}{80} + \tfrac{24}{80} = \tfrac{59}{80} \ (= .738)$$

There is an interesting and useful way of illustrating this procedure at a glance. This is done with the help of a diagram called a "tree diagram" (see Fig. 3.4). Starting at the extreme left, we form two "branches" corresponding to the two first stages (R_1 and G_1). Along these branches we put the probability of going along that branch. At the end of each branch we form two other branches, corresponding to the two possible second stages. Along each of these branches we put the *conditional probability* of proceeding along that branch, given that we have reached that juncture. Finally, each *path* from left to right represents an event whose probability may be computed by multiplying the probabilities along the branches. (This is the product rule, Equation 3.19.) In general, we do not even need "stages." At each juncture we have branches corresponding to mutually exclusive events. Then each path corresponds to the intersection of the events at the junctures.

In our example the event "final ball green" corresponds to the 2 paths terminating at G (darkened in Fig. 3.4), and we add the probabilities along these paths to find the probability of a path terminating at G.

The process may be repeated for more than 2 stages, because we may

3 However, see the remarks following Example 20.

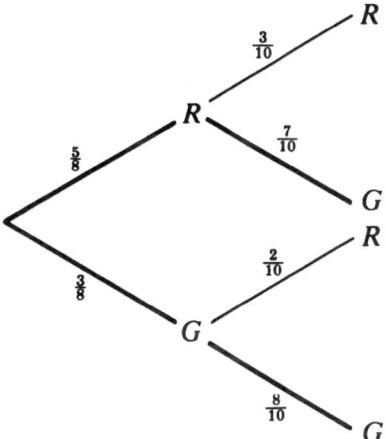

3.4   Tree Diagram for Urn Problem

apply the product rule repeatedly. We state and prove it for 3 events. The generalization to more than 3 events is straightforward:

$$p(A \cap B \cap C) = p(A) \cdot p(B|A) \cdot p(C|A \cap B) \qquad (3.20)$$

To prove this equation we write $p(A \cap B \cap C) = p((A \cap B) \cap C) = p(A \cap B) \cdot p(C|A \cap B)$. We then apply the product rule once again to $p(A \cap B)$ to obtain this equation. Diagramatically, we have the path of Fig. 3.5.

$$\xrightarrow{p(A)} A \xrightarrow{p(B|A)} B \xrightarrow{p(C|A \cap B)} C$$

3.5   Path for $A \cap B \cap C$

26   Example

Amy is happy 50 percent of the time, sad 20 percent of the time, and neutral 30 percent of the time. She is going shopping for a dress. When she is happy,

Happy $\xrightarrow{.9}$ Dress: .45
.50

.20 Sad $\xrightarrow{.1}$ Dress: .02

.30

Neutral $\xrightarrow{.4}$ Dress: .12
$$\overline{.59}$$

3.6   Buying a Dress

she will buy a dress with probability .9. When sad, she will buy one with probability .1. Otherwise, she will buy a dress with probability .4. What is the probability that she will buy a dress?

The entire analysis is summarized in Fig. 3.6. We do not include the branches where no dress is bought. The probability is 59 percent.

27 Example

Suppose that in Example 26, Amy comes home with a dress. What is the probability she was happy?

This is a conditional probability problem. We must find $p(\text{happy}|\text{dress})$. Using Equation 3.15 and referring to Fig. 3.6, we have

$$p(\text{happy}|\text{dress}) = \frac{p(\text{happy} \cap \text{dress})}{p(\text{dress})}$$

$$= \frac{.45}{.59} = \frac{45}{59} \, (=76.3\%)$$

Thus the purchase of a dress may be used as evidence of Amy's happiness.

We were given $p(\text{dress}|\text{happy})$ and we found $p(\text{happy}|\text{dress})$. The difference between the two is striking, both numerically and conceptually. We think of $p(\text{dress}|\text{happy})$, $p(\text{dress}|\text{sad})$, etc., as "before the fact" (a priori) conditional probabilities. The probabilities $p(\text{happy}|\text{dress})$, $p(\text{sad}|\text{dress})$, etc., are thought of as "after the fact" (a posteriori) probabilities. Thus, after we receive the information that the dress was bought, we may recompute the probability of "happy," given that this event occurred. The whole procedure is generalized in the following theorem.

28 Theorem. (Bayes' Theorem)

If the sample space S is the union of the mutually exclusive events $E_1, \ldots, E_s$, then

$$p(E_i|A) = \frac{p(E_i) \cdot p(A|E_i)}{p(E_1) \cdot p(A|E_1) + \cdots + p(E_s) \cdot p(A|E_s)} \tag{3.21}$$

To prove this formula, we use Equation 3.15:

$$p(E_i|A) = \frac{p(E_i \cap A)}{p(A)}$$

By the product rule, $p(E_i \cap A) = p(E_i) \cdot p(A|E_i)$. The event A is the union of the mutually exclusive events $A \cap E_1, \ldots, A \cap E_s$. (Thus A can happen in s mutually exclusive ways—with E_1, with $E_2, \ldots$, or with E_s.) Hence, by Theorem 3.10, $p(A) = p(A \cap E_1) + \cdots + p(A \cap E_s)$. Finally, we may apply the product rule to each of these summands to obtain Equation 3.21. The entire procedure is summarized in Fig. 3.7. In practice we may use a figure similar to this rather than memorize Equation 3.21.

$$E_1 \xrightarrow{p(A|E_1)} A : p(E_1 \cap A) = p(E_1) \cdot p(A|E_1)$$

$$E_i \xrightarrow{p(A|E_i)} A : p(E_i \cap A) = p(E_i) \cdot p(A|E_i)$$

$$E_s \xrightarrow{p(A|E_s)} A : p(E_s \cap A) = p(E_s) \cdot p(A|E_s)$$

with branch probabilities $p(E_1)$, $p(E_i)$, $p(E_s)$.

$$p(A) = p(E_1) \cdot p(A|E_1) + \cdots + p(E_s) \cdot p(A|E_s)$$

3.7 Bayes' Theorem

29 Example

The percentage of freshmen, sophomores, juniors, and seniors in Desolation High School is 30, 30, 20, and 20 percent, respectively. The probability that a freshman takes and passes mathematics is .90. Similarly, the probabilities for the other classes are, respectively, .80, .70, and .40. A student chosen at random at the end of the term has just taken and passed mathematics. Find the probability that he is a senior.

This is a clear-cut application of Bayes' theorem. The computation and method is given in Fig. 3.8. The probability that a student has taken and passed math is .73. The required probability is $.08/.73 = \frac{8}{73} (= .110)$.

Sr. $\xrightarrow{.4}$ math .08

Jr. $\xrightarrow{.7}$ math .14

Soph. $\xrightarrow{.8}$ math .24

Fr. $\xrightarrow{.9}$ math .27

.73

with branch probabilities .2, .2, .3, .3.

3.8

Finally, let us consider a "case-analysis" probability problem solved with the help of a tree diagram.

30 Example

Three cards are successively drawn from a deck. A player wins if the first card is an ace, or if the first 2 cards are pictures (J, Q, K), or if all 3 cards have the same suit. What is his probability of winning?

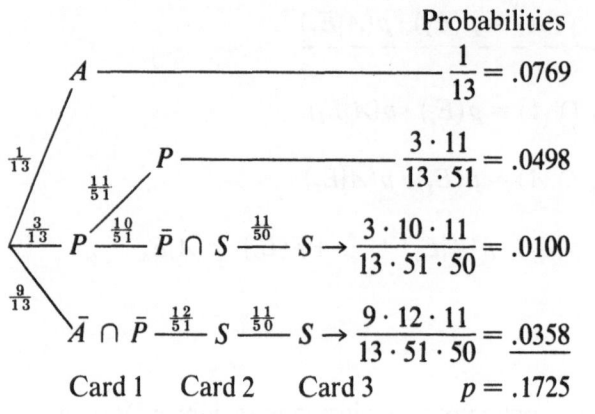

Probabilities

$$\frac{1}{13} = .0769$$

$$\frac{3 \cdot 11}{13 \cdot 51} = .0498$$

$$\frac{3 \cdot 10 \cdot 11}{13 \cdot 51 \cdot 50} = .0100$$

$$\frac{9 \cdot 12 \cdot 11}{13 \cdot 51 \cdot 50} = \underline{.0358}$$

Card 1   Card 2   Card 3 $p = .1725$

Code:
P: picture card
S: same suit
    as previous
A: Ace

3.9   Tree Diagram for Card Game

Figure 3.9 gives an appropriate tree diagram and solution. One word of caution is in order. We are using P for "picture card" and S for "same suit" in a generic way. These are not events. Thus the branch from P (card 1) to $\bar{P} \cap S$ (card 2) refers to the case that the first card is a picture and the second card is the same suit as the first but not a picture card. Here P stands for the event P_1 (first card is a picture card) and $\bar{P} \cap S$ stands for $\bar{P}_2 \cap S_{12}$ (second card is not a picture and the first 2 cards have the same suit). The probability $\frac{10}{51}$ along this branch is the conditional probability $p(\bar{P}_2 \cap S_{12}|P_1)$. Similar explanations are in order for all the other branches. In applications, we are often not too explicit about the underlying sample space or even the events in question.

EXERCISES

1. If, at a certain time, 60 percent of all television owners are watching television, and 32 percent of the watchers are watching a musical comedy, and of these only 20 percent are paying attention, what percentage of television owners are attentively watching a musical comedy?

2. Eighty percent of all the apples in a market are expensive. Ten percent of the expensive apples are tasteless. What percentage of the apples are tasteless and expensive?

3. A stamp collector orders 80 percent of his stamps from dealer X and 20 percent from dealer Y. He knows that X sends him a faulty stamp with probability .05, whereas Y sends him a faulty stamp with probability .15. What is the probability of receiving a faulty stamp? If a stamp is found to be faulty, what is the probability it came from dealer X?

4. Factories A, B, and C produce 30, 20, and 50 percent of a manufacturer's transistor radios. Quality control varies in these factories, so that 5, 7, and 2 percent of the output in factories A, B, and C are defective, respectively. A radio from a factory is discovered to be defective. Compute the probabilities that it came from factory A, from B, or from C.

5. In Example 25 suppose that the final ball is green. What is the probability that it was the ball from urn I?

6. Players A, B, and C are playing a chess tournament. Assume that A and B are equally matched and that both can beat C with probability .6. (Draw games are replayed.) In the tournament, A and B play each other, and then C plays the winner. Find the respective probabilities that A, B, or C wins the tournament. If the rules are changed so that A plays the winner of a B–C game, what is the probability of winning for each player?

7. The probability of skidding in the snow at a certain intersection is .1 if the car has snow tires but .4 is the car is not equipped with snow tires. A safety inspector, observing that intersection in the snow, discovers that 10 percent of the skidders have snow tires and the remaining skidders do not. The inspector wishes to estimate the percentage of cars in the city equipped with snow tires. Estimate this percentage, and state what assumptions you are using to derive this estimate.

8. Urn I contains 3 red and 7 black balls. Urn II contains 5 red and 5 black balls. They are identical in appearance. One urn is chosen at random and a ball is chosen from that urn. It is found to be black and is discarded. Another ball is chosen from that urn. What is the probability that it is black? What is the probability that urn I was chosen?

9. A bag contains 9 ordinary coins and 1 two-headed coin. A coin is chosen at random from that bag and is tossed 3 times, each toss yielding a head. What is the probability that the coin is two-headed? What is the probability that the next toss will be a head?

10. In Exercise 9 how many consecutive tosses of heads are necessary to deduce that the coin is two-headed with probability greater than .95?

11. When playing poker, a player discovers that Donald bluffs, at random, $\frac{1}{3}$ of the time. When bluffing he bets high 90 percent of the time and low the other 10 percent of the time. When not bluffing he bets high 50 percent of the time and low 50 percent of the time. Donald's opponent has observed that Donald is betting high. What is the probability that he is bluffing?

12. A mixture of grass seed contains 40 percent bluegrass, 55 percent ryegrass, and 5 percent crabgrass. On a certain lawn, the probability that bluegrass germinates is .4; the probability for ryegrass and crabgrass is .8

and .99, respectively. When this seed is used on that lawn, what percentage of the grass will be bluegrass? ryegrass? crabgrass?

13. Three cards are drawn from a deck. What is the probability of drawing 3 pictures, or 3 cards of the same suit, or a straight (ace high or ace low)?

14. An integer n is chosen at random between 1 and 5 inclusive. Once the number n is determined, an integer m is chosen at random between 1 and n inclusive. What is the probability that the second number is 1?

4 INDEPENDENCE

When we perform a series of probability experiments, we take care that the results of the experiments are *independent* of one another. Thus, if we choose one card from a deck to observe its color, we must make sure that when the experiment is repeated, we replace the card and shuffle well. If we were sloppy and put the card in the middle of the deck before proceeding to the new experiment, the probability of a particular color would *depend* on the color of the first card. We would be then dealing with conditional probabilities, because we know that the second card will not be the original card. We now formalize the notion of independent events. Roughly speaking, events A and B are independent if the probability of A is unchanged even if it is known that B happened.

31 Definition
Let S be a sample space and let A and B be events. We say that A and B are *independent* if

$$p(A) = p(A|B) \qquad (3.22)$$

We can put this equation into a more symmetric form. By Equation 3.15 we may write $p(A|B) = p(A \cap B)/p(B)$. Clearing fractions, we have

$$P(A \cap B) = p(A)p(B) \qquad (A \text{ and } B \text{ independent}) \qquad (3.23)$$

Conversely, we may divide this equation by $p(B)$ to obtain Equation 3.22. Equation 3.23 is, however, slightly more general than Equation 3.22 because it includes the case $p(B) = 0$, which was implicitly excluded in the original definition. We shall take the more general Equation 3.23 as the criterion for independence, thereby extending Definition 31 to include the case $p(B) = 0$. If $p(A \cap B) \neq p(A)p(B)$ we shall say that A and B are *dependent*.

The first observation we make is that the condition that A and B are independent is symmetric with respect to A and B. Thus, if A and B are independent, so are B and A. This follows immediately from Equation 3.23, because $A \cap B = B \cap A$. In the card example above, if we let $R_1 =$ red card on first experiment and $R_2 =$ red card on second experiment, then R_2

is independent of R_1, and R_1 is also independent of R_2. (The sample space is taken to be all ordered couples of cards with replacement.)

If Definition 31 is applied to a frequency table (interpreting relative frequencies as probabilities), we see that *dependent* events sometimes indicate a causal effect. In Table 3.2, suppose we let $M =$ unemployed males over 20 and $S =$ unemployed people 65 years and over. We have $f(M) = 65.6$ percent and $f(M|S) = 3.0/3.8 = 78.9$ percent. Thus a disproportionate number of males in the older category are unemployed. We can say that M and S are dependent, but we must be careful about saying that being over 65 tends to *cause* males to be unemployed. In fact, S and M are also dependent and we might also say that being over 65 and unemployed *causes* a person to be a male. The same argument raged (and will no doubt continue to rage) on whether cigarette smoking causes lung cancer. The undisputed evidence is that the event that a person is a smoker is dependent on the event that a person has lung cancer. It is certainly also theoretically conceivable that people prone to lung cancer happen to be more attracted to cigarettes, or that a common cause (perhaps nervousness, intelligence, etc.) is the reason for that dependence. Causal effects are difficult to demonstrate, but dependence of events can often be easily demonstrated. The intelligent person distinguishes between the two concepts.

We may generalize Definition 31 and Equation 3.23 to independence of several events $A_1, \ldots, A_s$.

32 Definition

The events $A_1, \ldots, A_s$ are said to be independent if

$$
\begin{aligned}
p(A_i \cap A_j) &= p(A_i)p(A_j) & 1 \leqslant i < j \leqslant s \\
p(A_i \cap A_j \cap A_k) &= p(A_i)p(A_j)p(A_k) & 1 \leqslant i < j < k \leqslant s
\end{aligned} \tag{3.24}
$$
$$
\vdots
$$
$$
p(A_1 \cap \cdots \cap A_s) = p(A_1) \cdots p(A_s)
$$

The first series of equations (the events taken 2 at a time) implies that any 2 of the events are independent. The second series, together with the first, shows that any event is independent of the simultaneous occurrence of 2 of the others: $p(A_i|A_j \cap A_k) = p(A_i)$ for distinct i, j, and k. (Use Equation 3.20 for a proof.) Similar interpretations are possible for all these equations. In all, there are $\binom{n}{2}$ equations using 2 different sets, $\binom{n}{3}$ equations using 3 different sets, etc. Thus there are $\binom{n}{2} + \binom{n}{3} + \cdots + \binom{n}{n}$ equations in all. By Equation 2.16, this may be simplified to $2^n - n - 1$. Thus we have $2^n - n - 1$ equations that define independence. For $n = 2$ we had $2^2 - 2 - 1 = 1$ equation. For $n = 3$ there are $2^3 - 3 - 1 = 4$ equations.

33 Example

Two dice are tossed. Let events A, B, and C be defined as follows: $A =$ "sum is 7," $B =$ "first die is 6," $C =$ "second die is 4 or over." We may compute $p(A) = \frac{1}{6}, p(B) = \frac{1}{6}, p(C) = \frac{1}{2}, p(A \cap B) = \frac{1}{36}, p(B \cap C) = \frac{1}{12}, p(A \cap C) = \frac{1}{12}$, $p(A \cap B \cap C) = 0$. Thus

$$p(A \cap B) = p(A)p(B)$$
$$p(A \cap C) = p(A)p(C)$$
$$p(B \cap C) = p(B)p(C)$$

but

$$p(A \cap B \cap C) \neq p(A)p(B)p(C)$$

Thus these 3 events are dependent, but any 2 of them are independent.

Independent events can occur quite abstractly. For example, consider the sample space given by the table

s	x	y	z	u
p	.2	.2	.3	.3

If $A = \{x, z\}$, $B = \{x, y\}$, then $A \cap B = \{x\}$. We have $p(A) = p(x) + p(z) = .2 + .3 = .5$, $p(B) = .2 + .2 = .4$, and $p(A \cap B) = .2$. Hence, since $.2 = .5 \times .4$, we have $p(A \cap B) = p(A)p(B)$. The events A and B are independent.

34 Example

A die is tossed repeatedly until a 6 turns up. What is the probability p_k that the 6 first occurs at the kth throw ($k = 1, 2, \ldots$)?

We note that we have an infinite sample space. We first reduce to a finite space by choosing a large number N and restricting ourselves to at most N tosses. In practice, $N = 100$ gives an excellent chance that a 6 will turn up. If we let $A_k =$ the event that a 6 occurs on the kth trial, the problem calls for the probability of $p(\bar{A}_1 \cap \bar{A}_2 \cap \cdots \cap \bar{A}_{k-1} \cap A_k)$ for $k = 1, 2, \ldots, N$. We make the following reasonable assumptions:

1. The events $\bar{A}_1, \bar{A}_2, \ldots, \bar{A}_{k-1}, A_k$ are independent. (This is our way of interpreting the notion that what happened on any of the previous trials does not affect the probability on a given trial.)

2. $p(A_i) = \frac{1}{6}$. (Thus, on any one trial, the 6 outcomes 1 through 6 are equally likely.)

Thus, by Equation 3.24, we have $p_k = p(\bar{A}_1 \cap \cdots \cap \bar{A}_{k-1} \cap A_k) = \frac{5}{6} \cdot \frac{5}{6} \cdots \frac{5}{6} \cdot \frac{1}{6} = \frac{1}{6}(\frac{5}{6})^{k-1}$ ($k = 1, 2, \ldots, N$). The probability that no 6 occurs during the N trials is $(\frac{5}{6})^N$. (For $N = 100$ this probability is 1.2×10^{-8}.) We also note that $p(\bar{A}_1 \cap \cdots \cap \bar{A}_{k-1} \cap A_k)$ is independent of N for $k \leq N$. It is therefore reasonable to take $\frac{1}{6}(\frac{5}{6})^{k-1}$ as the required answer. These probabilities are given in Table 1.6 to 3 decimal places. In that table, we chose $N = 12$, so there was a reasonable chance that no 6 would occur: $(\frac{5}{6})^{12} = .112$.

The results obtained in this example could be computed using the uniform probability space $A \times \cdots \times A$ (N factors), where $A = \{1, 2, \ldots, 6\}$. However, the present method is preferable for at least 2 reasons: First, it involves a simple and direct calculation. Second, assumptions 1 and 2 above give a simple mathematical translation of the underlying assumptions. A simple computation which also goes to the heart of the problem is always a happy occasion.

The following theorem gives some of the algebraic properties of independent events.

35 Theorem

(a) If A and B are independent, then so are $\bar{A}$ and B. More generally, if $A_1, \ldots, A_s$ are independent, so are $\bar{A}_1, A_2, \ldots, A_s$. (b) If A, B, and C are independent, then $A \cap B$ and C are independent, as are $A \cup B$ and C. More generally, if $A_1, \ldots, A_s$ are independent, $A_1 \cap A_2, A_3, \ldots, A_s$ and $A_1 \cup A_2$, $A_3, \ldots, A_s$ are independent.

To prove part (a), suppose A and B are independent. In general, $\bar{A} \cap B = B - (A \cap B)$. (The points in B and not in A are the points of B less the points of A that are in B. See Fig. 3.10.) Also $(A \cap B) \subseteq B$. Thus, using Equations 3.4, 3.23, and 3.2, we obtain

$$
\begin{aligned}
p(\bar{A} \cap B) &= p(B - (A \cap B)) = p(B) - p(A \cap B) \\
&= p(B) - p(A)p(B) \\
&= [1 - p(A)]p(B) \\
&= p(\bar{A})p(B)
\end{aligned}
$$

This proves that $\bar{A}$ and B are independent. The proof for independent events $A_1, \ldots, A_s$ is similar and is left to the reader.

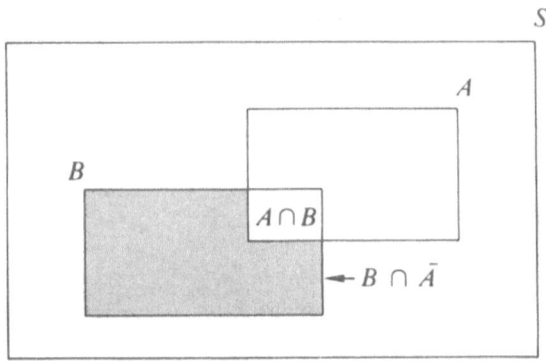

3.10 $B \cap \bar{A} = B - (A \cap B)$

To prove the result on intersections and union, we note that if A, B, and C are independent, then clearly $A \cap B$ and C are independent, because $p((A \cap B) \cap C) = p(A)p(B)p(C) = p(A \cap B)p(C)$. We may prove that $A \cup B$ and C are independent, by applying these results on complements on intersections, as well as De Morgan's laws (Equation 3.6). Thus we have

If A, B, C are independent, then $\bar{A}, B, C$ are independent.
Hence $\bar{A}, \bar{B}, C$ are independent.
Hence $\underline{\bar{A} \cap \bar{B}}, C$ are independent.
Hence $\overline{\bar{A} \cap \bar{B}}, C$ are independent.
Hence $A \cup B, C$ are independent.

The last step used the obvious formula $\bar{\bar{A}} = A$ and De Morgan's laws. The result for s independent events is proved similarly.

We may apply this theorem several times to obtain a result such as: If A, B, C, D, and E are independent, then $\bar{A} \cup B, \bar{D} \cap E$, and C are independent.

36 Example

Six dice are tossed. What is the probability that at least one 1 appears? How many dice are required to assure a better-than-even chance that at least one 1 appears?

If A is the event that at least one 1 appears, it is easier to consider the complementary event $\bar{A}$ that no 1 appears. This event equals the intersection of the event $\bar{A}_1$ (no 1 on the first die), $\bar{A}_2$ (no 1 on the second try), etc. Thus, since the A_i are independent, $p(\bar{A}) = p(\bar{A}_1 \cap \cdots \cap \bar{A}_6) = (\frac{5}{6})^6 \, (= .375)$. The required probability is therefore $p(A) = 1 - p(\bar{A}) = .625$. For the second part, we compute that $(\frac{5}{6})^3 = .58$, while $(\frac{5}{6})^4 = .48$. Thus 4 dice are required if a better-than-even chance is desired.

In this problem A_i was the event "1 on the ith die," and we wished to compute $p(A)$, where $A = A_1 \cup \cdots \cup A_6$. The complement was $\bar{A} = \bar{A}_1 \cap \cdots \cap \bar{A}_6$. This is a general situation (complements take unions into intersections and vice versa) and was already mentioned in Equation 3.6. The more general equations are also called De Morgan's laws:

$$\overline{A_1 \cup \cdots \cup A_s} = \bar{A}_1 \cap \cdots \cap \bar{A}_s \qquad \overline{A_1 \cap \cdots \cap A_s} = \bar{A}_1 \cup \cdots \cup \bar{A}_s$$
$$(3.25)$$

The proofs are left to the reader.

The procedure in Example 36 may be generalized.

37 Theorem

If $A_1, \ldots, A_s$ are independent events, with probabilities $p_1, \ldots, p_s$, respectively, then

$$p(A_1 \cup A_2 \cup \cdots \cup A_s) = 1 - (1 - p_1)(1 - p_2) \cdots (1 - p_s) \quad (3.26)$$

For example,

$$p(A_1 \cup A_2) = 1 - (1-p_1)(1-p_2) = 1 - (1-p_1-p_2+p_1p_2)$$
$$= p_1 + p_2 - p_1p_2$$

which is the form of Equation 3.5 in this case, because $p(A_1 \cap A_2) = p(A_1)p(A_2)$.

To prove Equation 3.26 we first note that $\bar{A}_1, \ldots, \bar{A}_s$ are independent by Theorem 35. Hence

$$p(\bar{A}_1 \cap \cdots \cap \bar{A}_s) = p(\bar{A}_1) \cdots p(\bar{A}_s)$$
$$= (1-p_1) \cdots (1-p_s)$$

But, by Equation 3.25, $\bar{A}_1 \cap \cdots \cap \bar{A}_s = \overline{A_1 \cup \cdots \cup A_s}$. Thus

$$p(\overline{A_1 \cup \cdots \cup A_s}) = (1-p_1) \cdots (1-p_s)$$

Finally, we obtain the result using Theorem 5.

Another example, taken from the theory of numbers, illustrates the power of Theorem 35.

38 Example

How many integers from 1 through 120 inclusive have no factor (except 1) in common with 120?

We factor 120 into primes: $120 = 2^3 \cdot 3 \cdot 5$. We see that we are looking for integers not divisible by 2, 3, or 5. Let S be the (uniform) probability space of integers from 1 through 120. Let A_2 be the event "divisible by 2," and in general let A_i be the event "divisible by i." Then *if i is a divisor of* 120, $p(A_i) = 1/i$. To see this, the numbers divisible by i are precisely $1 \cdot i, 2 \cdot i, \ldots,$ $(120/i) \cdot i$. There are $120/i$ such numbers and 120 sample points. The quotient is $1/i$.

Now we claim that A_2, A_3, and A_5 are independent. To see this, note that $A_2 \cap A_3$ consists of those numbers divisible by 2 and 3 — hence by 6. Thus $A_2 \cap A_3 = A_6$, and $p(A_2 \cap A_3) = p(A_2)p(A_3)$ since $\frac{1}{6} = (\frac{1}{2})(\frac{1}{3})$. Similar calculations apply for $A_2 \cap A_5$, $A_3 \cap A_5$, and $A_2 \cap A_3 \cap A_5$. We are looking for $n(\bar{A}_2 \cap \bar{A}_3 \cap \bar{A}_5)$. Since A_2, A_3, and A_5 are independent (Theorem 35), we have

$$p(\bar{A}_2 \cap \bar{A}_3 \cap \bar{A}_5) = (1-\tfrac{1}{2})(1-\tfrac{1}{3})(1-\tfrac{1}{5})$$

Thus, since there are 120 sample points, we have, by Equation 1.15,

$$n(\bar{A}_2 \cap \bar{A}_3 \cap \bar{A}_5) = 120 \cdot \tfrac{1}{2} \cdot \tfrac{2}{3} \cdot \tfrac{4}{5} = 32$$

Thus, in contrast to our previous attempts at computing probabilities by counting, we have here counted by computing probabilities.

The general situation in number theory is as follows. We let $\varphi(n)$ be the number of integers from 1 through n which have no factor, except 1, in common with n. We let $p_1, \ldots, p_k$ be all the *distinct prime* factors of n. Then

$$\varphi(n) = n\left(1 - \frac{1}{p_1}\right) \cdots \left(1 - \frac{1}{p_k}\right) \tag{3.27}$$

EXERCISES

1. If a card is chosen from a deck, show that the events "red" and "ace" are independent events.

2. In Exercise 1 are the events "red," "ace," and "diamond or club" independent? Show your calculations.

3. Two dice are tossed. Determine whether the events "6 on first die" and "sum is 8" are independent.

4. If $p(A \cap B \cap C) = p(A)p(B)p(C)$, are A, B, and C independent? Give reasons. (*Hint:* Construct an appropriate probability space, with $A = \{x, a\}, B = \{x, b\}, C = \{x, c\},$ and $S = \{a, b, c, d, x\}$.)

5. If S is a uniform probability space with k elements, and if A and B are independent events, find a formula relating $n(A)$, $n(B)$, and $n(A \cap B)$. Generalize to more than 2 sets.

6. Prove: If $A_1, \ldots . A_s$ are independent, then $A_1, \ldots, A_{s-1}$ are independent.

7. Prove: If $A_1, \ldots, A_s$ are independent, then $\emptyset, A_1, \ldots, A_s$ are independent.

8. A coin has probability p of landing heads and probability $1 - p$ of landing tails. Assuming that $0 < p < 1$, prove that $p(\text{HT}|(\text{HT} \cup \text{TH})) = \frac{1}{2}$. If a coin is suspected to be unfair, this result, properly used, will convert a coin into a "fair" coin. Explain how this may be done.

9. One person tosses a coin, another picks a card, and a third throws 2 dice.

    a. What is the probability that a head, a red card, and the sum 8 occurs?
    b. What is the probability that at least one of these events occur? State what assumptions you are making to arrive at your answer.

10. Suppose that the table of a sample space is as follows:

s	a	b	c	d	e	f
p	.12	.12	.16	.18	.20	.22

Let $A = \{d, e, f\}$ and $B = \{b, c, e, f\}$. Show that A and B are independent.

11. Ten people, chosen at random, are in a room. What is the probability that at least one of them has April 11 as his birthday? (*Hint:* Ignore leap years, assume uniformity and independence, and use Equation 3.26 and the binomial theorem.)

12. Some people answer Exercise 11 by claiming that since the probability is 1/365 in each case, the answer is 10/365, because there are 10 people. Using the binomial theorem, explain why this argument gives the approximate answer in this case but gives a terrible answer if there are 200 people in the room.

13. An urn contains 9 red balls and 1 white ball. Twenty balls are chosen, one at a time, with replacement. What is the probability that the while ball will be chosen? Do not evaluate.

14. A coin is tossed repeatedly until a head occurs. What is the probability that k tosses ($k = 1, 2, 3, \ldots$) are required?

15. Write out Equation 3.26 explicitly for $s = 3$. Identify each one of the summands as a probability of an appropriate event.

16. Generalize Exercise 15 to an arbitrary number s.

17. How many integers are there between 1 and 840 inclusive which have no factor except 1 in common with 840?

18. How many integers are there between 1 and 12,000 inclusive which have a common factor, larger than 1, with 12,000?

19. Prove: If $A \subseteq B$ and if A and B are independent, then either A is impossible or B is certain.

20. Alex, Ben, and Carl independently attempt to solve a puzzle. The probabilities that they solve the puzzle are, respectively .8, .9, and .7. Find the probability that the puzzle will be solved. State your assumptions.

21. A die is tossed repeatedly, for not more than 10 times. What is the probability that a 2 occurs before a 3? (i.e., a 2 occurs on some trial, and a 3 did not occur on the previous trials.) What is the probability that neither a 2 nor a 3 occurs?

22. A die is tossed until a 2 or 3 turns up. What is the probability that a 2 occurs before a 3?

23. Using Formula 3.5, prove that if A, B, and C are independent, then $A \cup B$ and C are independent.

24. Prove: If A and C are independent, B and C are independent, and $A \cap B = \emptyset$, then $A \cup B$ and C are independent.

*5 CONSTRUCTION OF SAMPLE SPACES

In the preceding sections we introduced many different probability spaces to compute various probabilities. In this section we shall investigate some of the underlying assumptions behind our choice.

We first note that we have consistently treated $p(A|B)$ as a *probability*, although the definition (Equation 3.15) does not, by itself, give us a probability space. Equation 3.15 is sufficient motivation for the following definition.

39 Definition

Let S be a probability space, and let $B \subseteq S$ with $p(B) \neq 0$. The probability space $S|B$ (*read: S restricted to B*) is the probability space whose sample points are the same as S but whose *probabilities* $p(s|B)$ are defined by the formulas

$$p(s|B) = \frac{p(s)}{p(B)} \qquad \text{if } s \text{ is in } B \tag{3.28}$$

$$p(s|B) = 0 \qquad \text{if } s \text{ is not in } B \tag{3.29}$$

To verify that this is a probability space in the sense of Definition 6 of Chapter 1, we must verify Equation 1.5 for this probability. (Equation 1.4 is a trivial consequence of Equation 1.5, because each probability is non-negative.) We point out that the sum $\Sigma_{s \in S} \, p(s|B)$ can be computed only for s in B, because the other terms are zero by Equation 3.29. We leave the details of this proof to the reader.

Often $S|B$ is defined by taking B as the sample space and using Equation 3.28. This amounts to ignoring the points of $\bar{B}$. The distinction between the two notions is not too critical, because the ignored points have probability 0.

Once Definition 39 is accepted, we may define the probability (in the restricted probability space $S|B$) of an event A in the sense of Definition 12 of Chapter 1. It is not surprising, of course, that this probability is simply $p(A|B)$. We leave the formal proof to the reader. In turn, this implies that all the concepts, theorems, and methods of probability theory apply to conditional probability. For example, $p(\bar{A}|B) = 1 - p(A|B)$ and $p(A \cup B|C) = p(A|C) + p(B|C) - p(A \cap B|C)$. We may even form conditional probabilities in $S|B$, although this does not lead to anything new. (See Exercises 4 and 5 at the end of this section.)

In the discussion following Example 20 we observed that $p(A|B)$ can be evaluated if the probabilities of A, given certain instances of B are known. This result is governed by the following theorem.

40 Theorem

Suppose B is the union of mutually exclusive events $B_1, \ldots, B_s$, and that

$p_i = p(B_i) \neq 0$. Suppose $p(A|B_i) = p$, the same value for each i. Then $p(A|B) = p$.

Remark. We have already noted in the discussion following Example 3.21 that it is essential that the B_i be mutually exclusive.

To prove this theorem, we first note that $B \cap A$ is the union of the mutually exclusive events $B_i \cap A$ (see Fig. 3.11). Thus, if an outcome is in B and in A it must be in exactly one B_i and in A, and conversely. Then we have

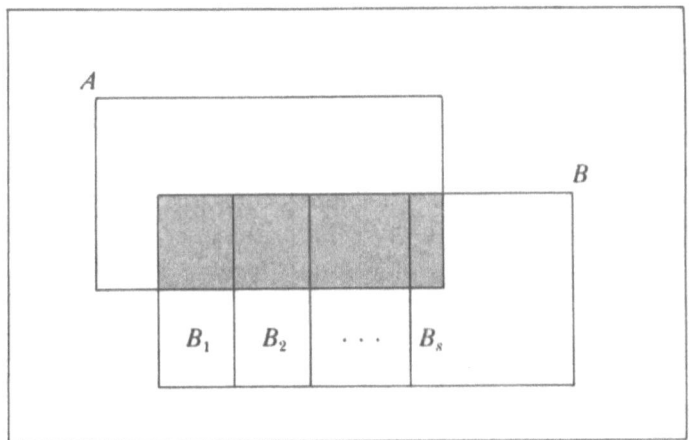

3.11 $(B_1 \cup \cdots \cup B_s) \cap A = (B_1 \cap A) \cup \cdots \cup (B_s \cap A)$

$$
\begin{aligned}
p(B \cap A) &= p(B_1 \cap A) + \cdots + p(B_s \cap A) \\
&= p(B_1) \cdot p(A|B_1) + \cdots + p(B_s) \cdot p(A|B_s) \\
&= p_1 p + \cdots + p_s p \\
&= (p_1 + \cdots + p_s)p = p(B)p
\end{aligned}
$$

because B is the union of the mutually exclusive B_i. Thus

$$
p(A|B) = \frac{p(A \cap B)}{p(B)} = p
$$

which is the result. The reader should construct a tree diagram to clarify the proof.

Another method of constructing probability spaces was to *identify* different outcomes by agreeing to regard them as the same outcome. In Example 1 of Chapter 1 we considered only 6 sample points, despite the common knowledge that there were more than 6 observable outcomes. Rather, we took the *event* A_6 "highest number is 6" and regarded it as one *sample point*, along with 5 other similarly constructed sample points. This procedure of identification may be defined as follows.

41 Definition

If S is a sample space with events $A_1, \ldots, A_r$ such that (a) the A_i are mutually exclusive and (b) $S = A_1 \cup \cdots \cup A_r$, then we say that the A_i's form a partition P of S.

42 Definition

If S is a sample space and $\{A_1, \ldots, A_r\}$ is a partition P of S, we define the new sample space $S \div P$ (*read: S* identified under *P*) by taking as sample points the events $A_1, \ldots, A_r$ and defining the probabilities $p(A_i)$ to be the probability of A_i as an event.

In brief, $p(A_i)$ is as it always was, but we now regard A_i as a sample point of $S \div P$.

We leave to the reader the easy verification that $S \div P$ is a sample space in the sense of Definition 8 of Chapter 1. A useful property of $S \div P$ is that the probabilities of events are consistent with probabilities in S. For example, if we take S as the sample space for 3 dice and A_i to be the event "high die is i," then $S \div P$ is the 6-point sample space of Example 1 of Chapter 1. The event E "high die is even" may be regarded in $S \div P$ as the event $\{A_2, A_4, A_6\}$ or as an event in S. In either case the probability is the same.

The following examples illustrate this identification procedure:

1. In any experiment, such as the 3-dice experiment of Example 1 of Chapter 1, when we identify an outcome by some characteristic, we are identifying in the sense of Definition 42. Each set A_i contains all the elements having a particular characteristic.

2. In D'Alembert's problem (Example 17 of Chapter 1) and in the dice-tossing problem of Example 34, we stopped the experiment as soon as an occurrence happened. Ordered n-tuples were required, and the procedure involved (in theory) continuing n times because "it didn't matter." In D'Alembert's problem we agreed that the event H was to be construed as $\{HT, HH\}$. In dice tossing, if we throw at most 3 times and if we stop at the first 6, we agree that the event $(1, 6)$ shall be regarded as a shorthand for $\{(1, 6, 1), (1, 6, 2), \ldots, (1, 6, 6)\}$. Thus $(1, 6)$ was an event A_{16} consisting of all triples $(1, 6, x)$. We regard A_{16} as a sample point because of this identification procedure.

3. When we go from ordered to unordered n-tuples, we identify different ordered n-tuples if they yield the same unordered n-tuple. This was done in poker. It can be done in tossing 2 identical dice and finding the sample space of observably different outcomes. Here $p\langle 1, 1 \rangle = \frac{1}{36}$, but $p\langle 1, 2 \rangle = \frac{2}{36}$, because $\langle 1, 2 \rangle$ is the event $\{(1, 2), (2, 1)\}$. (We use the symbol $\langle a, b \rangle$ to designate an unordered couple.)

Another, most important, construction is that of the product space $S \times T$, where S and T are probability spaces. We have already defined $S \times T$ as a set (see Definition 1 of Chapter 2), and we shall shortly define probabilities in $S \times T$. The motivation for this definition is that if we choose a point s of S independently of a point t of T, the probability of choosing s and t is $p(s)p(t)$. Choosing s and t may be regarded as choosing an ordered couple (s, t), and we are led to the following definition.

43 Definition
Let S and T be probability spaces. Then $S \times T$ is defined to be the probability space whose points are all ordered couples (s, t) with s in S and t in T and with probability defined by

$$p(s, t) = p(s)p(t) \tag{3.30}$$

44 Example
Let $S = \{a, b, c\}$ with $p(a) = .3, p(b) = .5, p(c) = .2$. Let $T = \{x, y, z\}$, with $p(x) = .1, p(y) = .7, p(z) = .2$. Then $S \times T$ has the 9 sample points of Fig. 3.12. The probabilities appear under the sample points. The probability of (b, z), for example, is given by Equation 3.30: $p(b, z) = p(b)p(z) = (.5)(.2) = .10$.

3.12 Product Space

T S	x .1	y .7	z .2
a .3	(a, x) .03	(a, y) .21	(a, z) .06
b .5	(b, x) .05	(b, y) .35	(b, z) .10
c .2	(c, x) .02	(c, y) .14	(c, z) .04

It is necessary to verify that $S \times T$ is, in fact, a probability space in the sense of Definition 6 of Chapter 1. We observe in Fig. 3.12 that the sum of each row in the body of the table equals the probability of the particular element in S which determines that row. We show this in general. Let $S = \{s_1, \ldots, s_k\}, T = \{t_1, \ldots, t_r\}$. Then for each s_i in S,

$$p(s_i, t_1) + p(s_i, t_2) + \cdots + p(s_i, t_r) = p(s_i)p(t_1) + p(s_i)p(t_2) + \cdots$$
$$+ p(s_i)p(t_r)$$
$$= p(s_i)[p(t_1) + p(t_2) + \cdots + p(t_r)]$$
$$= p(s_i)$$

Using the Σ sign, we have $\Sigma_{t \in T} p(s_i, t) = p(s_i)$. Thus, if we add the probabilities of all the elements in $S \times T$, we obtain (summing one row at a time) $p(s_1) + \cdots + p(s_k) = 1$. This proves Equation 1.4. As before, Equation 1.5 is trivial. Thus $S \times T$ is a sample space.

The extension to several spaces is straightforward. We merely define $S \times T \times R = (S \times T) \times R$, identifying the element $((s, t), r)$ with the ordered triple (s, t, r). Its probability is $p(s, t, r) = p(s)p(t)p(r)$. A similar procedure applies to n spaces.

45 Definition

If $S_1, S_2, \ldots, S_n$ are probability spaces, the probability space $S_1 \times S_2 \times \cdots \times S_n$ is the set of n-tuples $(s_1, \ldots, s_n)$, with s_i in S_i and with probability defined by

$$p(s_1, s_2, \ldots, s_n) = p(s_1)p(s_2) \cdots p(s_n) \qquad (3.31)$$

We may verify that this is a probability space by using the result for $n = 2$ over and over again.

An important special case is when the spaces S_i are identical. We then use the power notation.

46 Definition

$S^n = S \times S \times \cdots \times S$ (n factors).

Equation 3.31 shows that product spaces are the appropriate probability spaces to use if the experiments for $S_1, \ldots, S_n$ are to be performed independently. Thus each sequence of n possible outcomes in the n experiments is represented by *one* outcome $(s_1, \ldots, s_n)$ with probability given by Equation 3.31. *In particular, if one experiment (sample space S) is repeated, independently, n times, the appropriate sample space is S^n.* We have used this idea for coins and dice, always taking S^n to be uniform. The reason for the uniformity assumption can be explained using the following theorem.

47 Theorem

If S and T are uniform, so is $S \times T$.

In fact, if S has n elements, and T has m elements, $S \times T$ has nm elements. The probability of any one element is $p(s, t) = p(s)p(t) = (1/n)(1/m) = 1/nm$. Thus $S \times T$ is uniform.

The extension to n spaces and to S^n is immediate. Thus the assumption that a coin is fair ($S = \{H, T\}$ is uniform) and that the trials are independent implies that S^n is a uniform sample space.

If S and T are probability spaces and $A \subseteq S$, $B \subseteq T$, then $A \times B$ is an event in $S \times T$. (Verbally, $A \times B$ is the event "first component in A, second in B.") Since we think of A and B as independent, the following result is not surprising.

48   Theorem

If $A \subseteq S, B \subseteq T$, then

$$p(A \times B) = p(A)p(B) \tag{3.32}$$

To prove this theorem, write $A = \{a_1, \ldots, a_r\}$, $B = \{b_1, \ldots, b_s\}$. Fixing a_i in A, we have

$$p(\{a_i\} \times B) = p(a_i, b_1) + p(a_i, b_2) + \cdots + p(a_i, b_s)$$
$$= p(a_i)p(b_1) + p(a_i)p(b_2) + \cdots + p(a_i)p(b_s)$$
$$= p(a_i)[p(b_1) + \cdots + p(b_s)]$$
$$= p(a_i)p(B)$$

But A is the union of its elements: $A = \{a_1\} \cup \cdots \cup \{a_r\}$. Thus

$$p(A \times B) = p(\{a_1\} \times B) + \cdots + p(\{a_r\} \times B)$$
$$= p(a_1)p(B) + \cdots + p(a_r)p(B)$$
$$= [p(a_1) + \cdots + p(a_r)]p(B) = p(A)p(B)$$

This is the result. (Our previous remark that the sum of the probabilities in each row of $S \times T$ corresponding to an s in S is equal to the probability of s, is the special case $A = \{s\}, B = T$.)

This result immediately generalizes to n factors. The proof merely uses the result for two factors over and over.

49   Theorem

If $A_i \subseteq S_i$ $(i = 1, 2, \ldots, n)$, then

$$p(A_1 \times A_2 \times \cdots \times A_n) = p(A_1)p(A_2) \cdots p(A_n) \tag{3.33}$$

If A and B are events in S and T, respectively, it is not correct to say that A and B are independent events in $S \times T$. In fact, neither A nor B is an event in $S \times T$! However, we can say that the event $[A]_1$ (of $S \times T$) consisting of all elements whose first component is in A and the event $[B]_2$ of elements whose second component is in B, are independent.

50   Definition

Let $S = S_1 \times S_2 \times \cdots \times S_n$ be a product of n probability spaces, and let A be an event in S_i. We let $[A]_i$ be the set of all elements of S whose ith component is in A. Equivalently,

$$[A]_i = S_1 \times S_2 \times \cdots \times S_{i-1} \times A \times S_{i+1} \times \cdots \times S_n \tag{3.34}$$

In this case we say that the event $[A]_i$ is *determined* by A in the ith component.

It is an immediate consequence of Equation 3.33 that

$$p([A]_i) = p(A) \tag{3.35}$$

because all factors on the right-hand side of Equation 3.33 are 1, except $p(A)$. For example, if 7 dice are tossed, the probability that the third toss is a 4 or a 5 is $\frac{1}{3}$. Here $S = \{1, 2, 3, 4, 5, 6\}$, taken to be uniform, $A = \{4, 5\}$, and in S^7 we have $p([A]_3) = p(A) = \frac{2}{6} = \frac{1}{3}$. $[A]_3$ is the event of all possible throws on which the third die is a 4 or a 5.

The general notion that if the product space $S \times T$ is used, then any event of S is independent of any event of T, is expressed by the following theorem.

51   Theorem

In $S \times T$ suppose that $[A]_1$ is determined by A in the first component and $[B]_2$ is determined by B in the second component. Then $[A]_1$ and $[B]_2$ are independent.

To prove this we set $[A]_1 = A \times T$, $[B]_2 = S \times B$. Then $[A]_1 \cap [B]_2$ is the set of points whose first component is in A and whose second is in B: $[A]_1 \cap [B]_2 = A \times B$ (see Fig. 3.13). Thus

$$p([A]_1 \cap [B]_2) = p(A \times B) = p(A)p(B) = p([A]_1)p([B]_2)$$

This is the result.

3.13 $A \times B = [A]_1 \cap [B]_2$

In an entirely analogous way we arrive at the following useful generalization to n spaces.

52   Theorem

Let $S = S_1 \times \cdots \times S_n$, and let $A_1, A_2, \ldots, A_r$ be events in S, each determined by events in different components. Then the events $A_1, \ldots, A_r$ are independent.

In analogy with Theorem 11 or 11' of Chapter 1, it is possible to construct appropriate probability spaces for a many-stage experiment in which the occurrences (i.e., the sample space) at the ith stage depends upon what occurred at the previous stages. We shall not go into this in any detail, but the general idea is worth noting. For two stages $S \times T$ is made into a probability space by first assigning $p_{st} = p(2\text{nd component is } t | 1\text{st component is } s)$ ($=$ the probability of going from s to t), and then formally defining $p(s, t) = p(s)p_{st}$. (It is required that $\Sigma_{t \in T}\, p_{st} = 1$.) This device is appropriate, for example, in choosing ordered couples without replacement. We merely choose $p_{ss} = 0$. The generalization to several spaces is similar to the generalization of product spaces to several factors.

EXERCISES

1. Prove: $S \,|\, S = S$.

2. Show that in $S \,|\, B$, the ratio $p(s_1)/p(s_2)$ is preserved for s_1 and s_2 in B:

$$\frac{p(s_1)}{p(s_2)} = \frac{p(s_1|B)}{p(s_2|B)} \qquad \text{for } s_1, s_2 \text{ in } B$$

(For example, if s_1 is twice as likely as s_2 and both are in B, then s_1 remains twice as likely as s_2 if B is given.)

3. Let S be a probability space and let $B \subseteq S$ with $p(B) \neq 0$. Let $p'(s)$ be a new probability defined in S such that

a. $\dfrac{p'(s_1)}{p'(s_2)} = \dfrac{p(s_1)}{p(s_2)} \qquad$ for s_1, s_2 in B.

b. $p'(s) = 0 \qquad$ for s not in B.

Prove that $p'(s) = p(s|B)$. [*Remark:* Equation a means that $p'(s_1)p(s_2) = p(s_1)p'(s_2)$. This avoids all questions about dividing by 0.]

4. Prove: $(S\,|\,B)\,|\,C$ exists if and only if $p(B \cap C) \neq 0$.

5. Prove: $(S\,|\,B)\,|\,C = S\,|\,(B \cap C)$ if $p(B \cap C) \neq 0$.

6. Let S be a finite probability space and let Z be the set of points whose probability is 0. Show how $S - Z$ may be made into a probability space $\tilde{S}$ in a natural manner.

7. Using the notation of Exercise 6, prove that if S is uniform and $p(B) \neq 0$, then $(S\,|\,B)$ is uniform.

8. Let $S = \{a,\ b,\ c,\ d,\ e,\ f\}$, with $p(a) = p(b) = .2$, $p(c) = .3$, and $p(d) = p(e) = p(f) = .1$. Let $A = \{a, c, d\}$, $B = \{a, e, f\}$, $C = \{c, d, e, f\}$.

Construct the probability space for

    a. $S|A$ **b.** $S|(B \cap C)$
    c. $S \times S$ **d.** $S \div P$, where P is the partition $\{A, \bar{A}\}$
    e. $(S \times S)|(B \times C)$ **f.** $(S \times (S|B))|(S \times (S|C))$

9. Suppose that $B = B_1 \cup B_2$ with $B_1 \cap B_2 = \emptyset$. Given that $p(A|B_1) = \frac{1}{2}$ and $p(A|B_2) = \frac{1}{3}$, is it possible to compute $p(A|B)$? Explain. What can be said about the value of $p(A|B)$? If $p(B_1) = p(B_2)$, find $p(A|B)$. Suppose B_1 is twice as likely as B_2. Find $p(A|B)$ in this case.

10. The following experiments have probability spaces that can be written in terms of restrictions, identifications, and/or products. Illustrate how this may be done, and briefly discuss your underlying assumptions.

    a. Jim plays Jean 5 games of Ping-Pong, the play stopping when a player has won 3 games. The probability that Jim wins a game is .6.

    b. A die is tossed until a 5 or a 6 turns up, or for 10 tosses, whichever comes first.

    c. An urn contains 5 black balls, 8 white balls, and 13 green balls. A ball is chosen and the color is observed.

    d. In part c, 2 balls are chosen and the colors are observed.

    e. Two identical-looking dice are tossed, and the numbers that appear are recorded.

11. The probability space of a loaded die is given by the following table:

s	1	2	3	4	5	6
p	.2	.2	.1	.1	.2	.2

Two such dice are tossed. Construct the probability space for the outcomes. Find the probability of the sum 7. Of the sum 7 or 11. Find the probability that the high die is a 6. Find the probability that either a 6 or a 1 appears.

12. When shopping for a car, Jack will buy brand X, Y, or Z with probability .3, .5, and .2, respectively. The color of his car will be red, tan, or green with probability .7, .2, and .1, respectively. Assuming that color and brand are independent, construct an appropriate probability space for the descriptions of Jack's future car.

13. In Exercise 12 verify Equation 3.32 if $A = \{X, Y\}$ and $B = \{$tan, green$\}$.

14. If n dice are tossed $(n \geqslant 2)$, express the following events using the $[A]_i$ notation.

    a. The first die is a 4 or a 5.

    b. The first or second die is a 6.

    c. The first or last die is even.

Compute the probabilities of the events in parts a, b, and c.

15. The following table is a table of the probabilities in the product space $S \times T$. Fill in the missing entries. (*Hint:* In any product space, prove that any 2 rows are proportional.)

S \ T	t_1	t_2	t_3
s_1	.05	.02	.03
s_2		.08	
s_3			

16. Suppose a coin has probability p of landing heads and probability $q = 1 - p$ of landing tails. The coin is tossed 4 times. Construct the probability space. (*Hint:* Let $A = \{H, T\}$. Find A^2, then $A^2 \times A^2 = A^4$.)

17. An experiment has the probability space $S = \{a, b, c\}$, with $p(a) = .5$, $p(b) = .4$, $p(c) = .1$. The experiment is run for as many times as necessary to obtain the same result twice. Construct an appropriate probability space for the possible outcomes. (*Hint:* Use a tree diagram.)

18. Five cards are taken at random from a deck. Using Theorem 40, find the probability that 4 of the cards have the same rank, given that 3 have the same rank. Explain why this theorem cannot be used to find the probability that 3 have the same rank, given that 2 have the same rank.

*CHAPTER 4 MISCELLANEOUS TOPICS

*1 REPEATED TRIALS

Suppose that an experiment has probability space S, and that A is an event of S with probability $p = p(A)$. We let P_n denote the probability that the event A will occur at least once if the experiment is repeated, independently, for n times. By Theorem 37 of Chapter 3, we have

$$P_n = 1 - (1-p)^n \tag{4.1}$$

The cases $p = 0$ and $p = 1$ naturally give $P_n = 0$ and $P_n = 1$, while the case $n = 1$ naturally gives $P_1 = p$. We shall therefore usually assume that $n > 1$ and

$$0 < p < 1 \tag{4.2}$$

In this section we shall be concerned with *estimating* the size of P_n for large values of n. To obtain the proper orientation, the reader should imagine p small and n large. Thus the event A will probably not occur on a given trial, but we are repeating the experiment many times. These tendencies tend to have opposite effects: $1 - (1-p)^n$ is near 0 if p is very small but is near 1 if n is very large. (Why?) We shall soon see that the behavior of P_n for large n is largely dependent on the value np.

1 Example
Ten people are in a room. What is the probability that at least 1 person was born on June 1?

We make the simplifying assumption that the probability of being born on June 1 is $p = \frac{1}{365}$ and that the birthdays are independent. Thus

$$P_{10} = 1 - (1 - \tfrac{1}{365})^{10} = 1 - (\tfrac{364}{365})^{10}$$

If we compute P_{10} to 3 significant figures, we have $P_{10} = .0271$.

In this example $np = \frac{10}{365} = .0274$. The following theorem can be used to approximate P_n if np is small.

2 Theorem

$$P_n = 1 - (1 - p)^n \approx np \qquad \text{if } np \text{ is small} \tag{4.3}$$

More precisely,

$$np - \frac{(np)^2}{2} < P_n < np \tag{4.4}$$

In the above example, $np = .0274$. Thus $.0274 - (.0274)^2/2 < P_{10} < .0274$. Since $(.0274)^2/2 < (.03)^2/2 = .0009/2 < .0005$, we have the (crude) approximation $.0269 < P_n < .0274$. This is good enough to give $P_n = .027$ to 3 decimal places. In general, the smaller np is, the more effective is the estimate for P_n given by Equation 4.4.

Equation 4.3 may be made reasonable by using the binomial theorem. We have $(1 - p)^n \approx 1 - \binom{n}{1}p = 1 - np$ if np is small. Thus $P_n = 1 - (1 - p)^n \approx 1 - (1 - np) = np$ if np is small. We shall not prove Formula 4.4, because the proof would take us too far afield.

3 Example

If 500 people are in a room, what is the probability that at least 1 of these people has his birthday on June 1?

As in Example 2, we have $P_{500} = 1 - (1 - \frac{1}{365})^{500}$. Here $p = \frac{1}{365}$, $n = 500$, and $np = \frac{500}{365} = 1.370$, and the approximation formula 4.3 or 4.4 is no help. A calculation using logarithms yields $P_{500} = .746$.

It turns out that there is a very useful approximation for P_n when n is large, which we shall now indicate. We shall compute $(1 + x/n)^n$ by the binomial theorem and then we see that happens as n gets large. (In the above example $x = -500/365$ and $n = 500$.) We have

$$\left(1 + \frac{x}{n}\right)^n = 1 + n \cdot \frac{x}{n} + \frac{n(n-1)}{2!} \cdot \frac{x^2}{n^2} + \frac{n(n-1)(n-2)}{3!} \cdot \frac{x^3}{n^3}$$

$$+ \cdots + \frac{n!\, x^n}{n!\, n^n}$$

$$= 1 + x + \frac{(1 - 1/n)}{2!} x^2 + \frac{(1 - 1/n)(1 - 2/n)}{3!} x^3 + \cdots$$

This expression suggests that we consider the *infinite series*

$$1+x+\frac{x^2}{2!}+\frac{x^3}{3!}+\cdots$$

because, when n is large, $1/n \approx 0, 2/n \approx 0$, etc. This series is one of the most famous in higher mathematics. It is proved in calculus texts that there is a certain number $e = 2.71828\ldots$ (named after the Swiss mathematician Euler) and that this series is the series for e^x:

$$e^x = 1+x+\frac{x^2}{2!}+\frac{x^3}{3!}+\cdots \tag{4.5}$$

The letter e is always reserved for this constant in much the same way that π is reserved for the constant $3.14159\ldots$. A proof of Equation 4.5 is beyond the scope of this text. Appendixes A and B give values for e^x for different values of x, which we shall freely use. From the above analysis it seems reasonable to expect the approximation

$$\left(1+\frac{x}{n}\right)^n \approx e^x \qquad \text{for large } n \tag{4.6}$$

Equivalently, setting $x/n = y$,

$$(1+y)^n \approx e^{ny} \qquad \text{for large } n \text{ and small } y \tag{4.7}$$

We state without proof a more precise formulation of these approximations which are analogous to Equation 4.4.

4 Theorem. Approximation of $(1+x/n)^n$

$$\left(1-\frac{x^2}{2n}\right)e^x < \left(1+\frac{x}{n}\right)^n < e^x \qquad (x > 0) \tag{4.8}$$

$$\left(1-\frac{x^2}{n}\right)e^{-x} < \left(1-\frac{x}{n}\right)^n < e^{-x} \qquad \left(0 < x < \frac{n}{2}\right) \tag{4.9}$$

In particular, if we call $x = \mu = np$ in Equation 4.9, we obtain the approximation

$$\left(1-\frac{\mu^2}{n}\right)e^{-\mu} < (1-p)^n < e^{-\mu} \qquad (\mu = np, 0 < p < \tfrac{1}{2}) \tag{4.10}$$

In Example 3 we had $n = 500$ and $p = \frac{1}{365}$. Thus $\mu = np = \frac{500}{365} = 1.37$. From Appendix A we find that $e^{-1.37} = .254$. The inequality (4.10) shows that $(1-p)^n$ is less than .254 but greater than $.254 - (1.37)^2/500 \times .254 \approx .253$. Thus $P_{500} = 1 - (1-p)^n \approx 1 - e^{-\mu} = 1 - .254 = .746$, in agreement with the value already given. Using Equation 4.10, we have

If $p < \frac{1}{2}$ and $\mu = np$, then

$$P_n \approx 1 - e^{-\mu} \tag{4.11}$$

with error at most $(\mu^2/n)e^{-\mu}$. In all cases, $P_n > 1 - e^{-\mu}$.

5 Example

A tax auditor decides that he will audit any tax return whose tax bill ends with 33 cents. He has 200 returns. What is the probability he will audit at least one of these?

It is reasonable to assume independence and to take $p = .01 =$ probability of a return with a tax bill ending with 33 cents. Thus $n = 200$, $p = .01$, and $\mu = np = 2$. By Appendix A, $e^{-\mu} = e^{-2} = .135$. Thus $P_{200} \approx 1 - .135 = .865 = 86.5$ percent. According to the inequality 4.10, P_{200} is larger than .865 with an error of at most $e^{-\mu}(\mu^2/n) = (.135)(4/200) < .003$.

6 Example

If an experiment has probability .001 of succeeding, how many times must this experiment be scheduled in order to ensure that it succeeds at least once, with probability larger than .95?

We have $P_n > 1 - e^{-\mu}$. Thus we may ensure $P_n > .95$ by making $1 - e^{-\mu} > 95$. This is equivalent to $e^{-\mu} < .05$. Refering to Appendix A, we have $e^{-3.00} < .05$. Thus we choose $\mu = 3.00$. Also, we have $p = .001$. Thus $\mu = np$ implies that $3.00 = n(.001)$ and $n = 3000$.

If $\mu = np$ is large, the inequality $(1-p)^n < e^{-\mu}$ shows that $(1-p)^n$ is very small, and hence $P_n = 1 - (1-p)^n$ is very close to 1. To see how this operates, the cases $\mu = 4$, 5, and 6 yields $e^{-\mu} = .018$, .006, and .002, respectively, so $P_n > .982$, .994, and .998, respectively. For $\mu = 10$, $e^{-\mu} = .0000454$, and $P_n > .9999546 = 99.99546$ percent. Thus, if we have 1 chance in 1 million of succeeding $(p = 10^{-6})$, but if we try 10 million (independent) times, $(n = 10^7)$, then it is virtually certain that we will succeed. $(\mu = 10, P_n > .99995.)$

We are thus transported back to Chapter 1, where we imagined a large number of experiments and where we expected the relative frequency of an event A to be near the probability $p = p(A)$. In particular, if $p > 0$, we expected, at any rate, that the event A would *occur* if we had a large number of experiments. We shall now make this more formal. Since we do not wish the analysis to depend on the function e^x, we shall give a direct and simple proof of this result.

7 Theorem

Let an event A have probability $p > 0$. In n independent trials, the probability P_n that A occurs at least once satisfies the inequality

$$1 - \frac{1}{np} < P_n < 1 \tag{4.12}$$

Also if ϵ is any positive number, then for some positive integer N,

$$1 - \epsilon < P_n < 1 \qquad \text{if } n > N \tag{4.13}$$

To prove this result, we shall estimate $(1-p)^n$. Since $(1-p)^n(1+p)^n = (1-p^2)^n < 1$, we have

$$(1-p)^n < \frac{1}{(1+p)^n} = \frac{1}{1+np+\binom{n}{2}p^2+\cdots} < \frac{1}{np}$$

In this latter inequality we have ignored all terms except one, thereby obtaining a smaller denominator and larger answer. Thus $1 > P_n = 1-(1-p)^n > 1-1/np$. Also $1/np < \epsilon$ if $n > N = 1/p\epsilon$. In this case $P_n > 1-1/np > 1-\epsilon$. This completes the proof.

The inequality 4.12 is extremely weak. For example, if $\mu = np = 10$, this inequality yields $P_n > 1-\frac{1}{10} = .9$, whereas we actually have $P_n > 1-e^{-10} = .99995$. The weakness lies in the proof of Equation 4.12, where we replaced $(1+p)^n$ by np. Still, the inequalities 4.12 and 4.13 are strong enough to show that for fixed $p > 0$, if n is large enough, P_n will be very close to 1, and in fact will be as close to 1 as we want provided that n is large enough.

The above results may be stated using the language of limits. We may state: If $0 < p \leqslant 1$, then $(1-p)^n \to 0$ as $n \to \infty$, and $P_n = 1-(1-p)^n \to 1$ as $n \to \infty$. (The symbol "$\to$" is read "approaches.") Theorem 7 may be paraphrased in the following interesting way.

7′ Theorem

If an event A has probability $p > 0$, and if the experiment is repeated indefinitely, then A will occur with probability 1.

This statement is understood to be another way of stating that $P_n \to 1$ as $n \to \infty$.

EXERCISES

Most of the exercises below need the values in Appendix A or B for numerical answers.

1. When playing poker, the probability of picking up a flush or better is .00366. During one evening, Harold will play 100 games of poker. Approximately what is the probability that he will pick up a flush or better during the evening? Use Equations 4.3 and 4.11 to obtain an approximation. Estimate how large your error is in each case.

2. Each person in a class of 20 people is told to toss 3 dice 10 times and record the sum. The teacher is reasonably sure that some one will have tossed the sum 3 or 4. How sure is he of this? What is the probability that someone reports the sum 18?

3. How many times should a pair of dice be tossed in order that a double 6 will occur with probability larger than $\frac{1}{2}$?

4. A printer will print a perfect page with probability .96. What is the probability of his printing a perfect 100-page book?

5. A certain bank sends out all its statements automatically, using a machine. The machine will send out a wrong statement with probability .0007. The bank sends out 800 statements. What is the probability that a wrong statement was sent out?

6. Using the tables of e^x and e^{-x}, compute approximately:
   a. $(1.03)^{30}$ **b.** $(.996)^{200}$
   c. $(.999)^{1,200}$ **d.** $(1.000123)^{9,000}$
In each case, estimate the error.

7. In playing roulette, the probability of winning on any one play is $\frac{1}{38}$. Approximately what is the probability of winning at least once if 4 games are played? If 25 games are played?

8. How many games of roulette (see Exercise 7) should a person plan to play in order to win at least once with probability greater than .9?

9. One hundred and twenty-five people each choose a number from 1 through 100 at random. What is the probability that the number 72 was chosen?

10. Last year there were about 1,500,000 marriages in the United States. Estimate the probability that for at least one of these couples, both partners had birthdates on June 24? Estimate the probability that for at least one of these couples, both partners had birthdays on February 29. State what assumptions are involved and if they seem reasonable.

11. Ten dice are tossed. Estimate the probability that at least one 6 occurs. Give an upper and lower bound for the probability.

12. If a person has probability .0001 of winning a certain lottery, and if he buys 1 ticket for 50 consecutive years, estimate his chances of winning during these 50 years.

13. A person picks up 5 cards from a deck and finds that they are {5 He, 6 Di, 9 He, J Cl, K Di}. He claims that a very rare event occurred, because the probability of drawing this hand is $1/\binom{52}{5} \approx .0000004$. Yet no one is very surprised at this occurrence. Explain why not. Your explanation should also explain why most people will be impressed if his hand was {10 Sp, J Sp, Q Sp, K Sp, A Sp}.

14. Each person in a class of 100 people tosses 10 coins. Approximately what is the probability that at least 1 person tosses 10 heads or 10 tails? What is the probability that at least 1 person has a $10-0$ or a $9-1$ distribution of heads and tails?

15. Using Equation 4.5 find the value of
  a. $e^{.1}$ **b.** $e^{-.1}$
  c. $e = e^{1.0}$ **d.** $1/e = e^{-1.0}$
Compare with the values in Appendixes A and B.

*2   INFINITE PROCESSES

If a sample space or an event contains infinitely many outcomes, many of our previous techniques are not applicable. However, by approximating by finite sets, and by taking limits, it is possible to obtain a satisfactory notion of probability for these events. For example, if an event has probability $p > 0$ of occurring, we have considered in Section 1 the probability that the outcome will eventually occur if the experiment is repeated indefinitely. Our method was to consider P_n, the probability that the event occurs at least once during n trials. If the original finite sample space was S, we thus considered S^n (a finite space), and we found that $P_n = 1 - (1-p)^n$. Finally, the probability of eventual occurrence was taken to be 1, because $P_n \to 1$ as $n \to \infty$. (See Theorem 7 or 7′.) This is a reasonable answer, because, for example, we may arrange to have $P_n > .99999$ by choosing n large enough. More generally, for any $\epsilon > 0$, no matter how small, we have $1 - \epsilon < P_n < 1$, provided that n is large enough.

In this section we shall consider several examples involving infinite processes, and we shall find probabilities by using analogous limiting techniques. Before doing so, we state without proof some useful results about limits.

8   Definition

Let a_n be a number for every positive integer n. We say that $a_n \to a$ (read: a_n approaches a) as $n \to \infty$, or that the limit of a_n is a if, for any $\epsilon > 0$, there is a number N with the property that for all $n > N$, we have $a - \epsilon < a_n < a + \epsilon$.

Briefly, for large enough n, a_n is within the range $a \pm \epsilon$. Informally, $a_n \approx a$ for large n. The number ϵ is the accuracy with which a_n approximates a and it may be arbitrarily prescribed as long as it is positive. The number N (depending on ϵ) tells us how far out in the sequence we must go before we are within ϵ and stay within ϵ, of the limit a.

9   Theorem. On Limits

If $a_n \to a$ and $b_n \to b$ as $n \to \infty$, then

$$a_n + b_n \to a + b \tag{4.14}$$

$$a_n - b_n \to a - b \tag{4.15}$$

$$a_n b_n \to ab \tag{4.16}$$

$$a_n/b_n \to a/b \qquad (\text{if } b_n \neq 0 \text{ and } b \neq 0) \tag{4.17}$$

$$ca_n \to ca \qquad (\text{for any constant } c) \tag{4.18}$$

The following result is useful and intuitively clear.

10   Theorem

If a_n is increasing with n, and if $a_n \leqslant K$ for fixed K and all n, then $a_n \to a$ for some value $a \leqslant K$. Similarly, if a_n decreases with n, and if $a_n \geqslant L$, then $a_n \to a$ for some $a \geqslant L$.

In this theorem the word "increasing" means "$a_n \leqslant a_{n+1}$ for all n." Thus we permit some or all of the terms to be stationary. Similarly, the word "decreasing" means "$a_n \geqslant a_{n+1}$ for all n."

In many of our applications a_n will be a probability. Thus $0 \leqslant a_n \leqslant 1$ and we may choose $K = 1$ or $L = 0$. The limit a will therefore also satisfy $0 \leqslant a \leqslant 1$.

Finally, we may restate the result proved in Theorem 7 and generalize it to include some negative values.

11   Theorem

If $-1 < a < 1$, then $a^n \to 0$.

We recall the result of algebra concerning the *infinite geometric series* with ratio r:

$$a + ar + ar^2 + \cdots + ar^{n-1} + \cdots = \frac{a}{1-r} \qquad \text{if } -1 < r < 1 \tag{4.19}$$

By this we mean that the sum of n terms approaches $a/(1-r)$ as $n \to \infty$. In fact,

$$a + ar + \cdots + ar^{n-1} = \frac{a - ar^n}{1-r} \tag{4.20}$$

and we may take limits using Theorem 11. Since $r^n \to 0$, the limit is $a/(1-r)$.

We now show how some of these results can be used to find probabilities.

12   Example

Alex and Bill alternately toss a coin. The first player to toss a head wins. Alex goes first. What is each player's probability of winning?

We shall give 2 methods for solving this problem.

Explicit Solution.   Alex wins if the event $A = \{H, TTH, TTTTH, \ldots\}$ occurs. Thus the probability that Alex wins is $p(A) = \frac{1}{2} + \frac{1}{8} + \frac{1}{32} + \cdots$. This is an infinite series with ratio $r = \frac{1}{4}$. By Equation 4.19 we have

$$p(A) = \frac{\frac{1}{2}}{1 - \frac{1}{4}} = \frac{\frac{1}{2}}{\frac{3}{4}} = \frac{2}{3}$$

In the same way, Bill wins if the event $B = \{TH, TTTH, \ldots\}$ occurs. This probability is $\frac{1}{4} + \frac{1}{16} + \cdots = \frac{1}{3}$, again by Equation 4.19.

The use of infinite series to evaluate this probability suggests that a limiting probability is involved. This may be seen as follows. If we limit the game to at most n tosses for Alex, the probability that he wins can be seen to be the *finite* sum $\frac{1}{2} + \frac{1}{8} + \cdots + 1/2 \cdot 4^{n-1}$. The infinite series is the limit of this sum as $n \to \infty$.

Indirect Solution.   Let $x =$ probability that Alex wins when it is his turn to toss, and let $y =$ probability that Alex wins when it is Bill's turn to toss.

Then, when Alex tosses, he can win by tossing H (probability $\frac{1}{2}$), or by tossing T (probability $\frac{1}{2}$) and then winning (probability y). Thus $x = \frac{1}{2} + \frac{1}{2}y$. Similarly, if it is Bill's turn, Alex can win by Bill's tossing a tail (probability $\frac{1}{2}$) and then winning on his turn (probability x). Thus $y = \frac{1}{2}x$. (See Fig. 4.1 for an appropriate tree diagram for these equations.) Thus we have

$$x = \tfrac{1}{2} + \tfrac{1}{2}y \qquad y = \tfrac{1}{2}x \qquad\qquad (4.21)$$

Solving simultaneously, we obtain $x = \frac{2}{3}$, $y = \frac{1}{3}$. Thus Alex's chances of winning is $\frac{2}{3}$. In a similar way, we may find that Bill's chance is $\frac{1}{3}$.

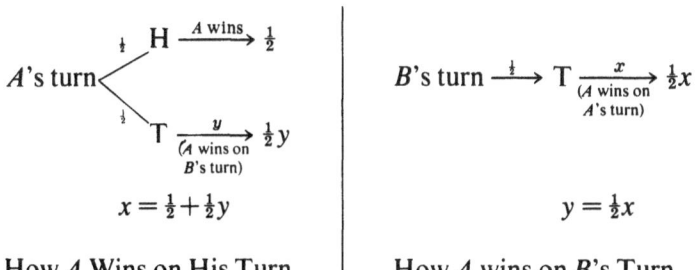

How A Wins on His Turn   How A wins on B's Turn

4.1   Tree Diagram for Coin-Tossing Game

Here x and y are also limiting probabilities. To see this, suppose we restrict the game to n tosses. Then we have x_n and y_n as the probability that Alex wins if it is his or Bill's toss, respectively. Now, once a player tosses a coin and a tail occurs, they are in the $(n-1)$ toss game. Thus we have

$$x_n = \tfrac{1}{2} + \tfrac{1}{2}y_{n-1} \qquad y_n = \tfrac{1}{2}x_{n-1} \qquad\qquad (4.22)$$

Now, it seems clear that increasing n cannot decrease Alex's chances of winning the game. Thus x_n and y_n are increasing, and we are assured that x_n and y_n have limits by Theorem 10: $x_n \to x$, $y_n \to y$. Finally, taking limits in Equation 4.22, we arrive at Equation 4.21. The *infinite* game was easier to handle, because at any unfinished stage of the game, the remaining game was always one of two possibilities (Alex's or Bill's turn) and did not depend upon how many tosses were left. In similar problems we shall use this infinite-game technique. The justification will be similar to the above remarks.

13 Example

Players A and B play a game in which each has probability $\frac{1}{2}$ of winning. When any player wins, he wins a penny from the other. They keep on playing until one of the players has no coins. Assuming that A starts with 3 cents and B with 7, what are their respective chances of winning everything?

If we consider the number n of coins which A has, we see that at any stage, n changes to $n-1$ or $n+1$, each with probability $\frac{1}{2}$. This continues until $n = 0$ or 10, at which point the game terminates. We visualize this as a *random walk*. At any stage, we toss a coin to determine whether we go left or right for one unit. This continues until we reach the edge ($n = 0$ or $n = 10$). (See Fig. 4.2.)

$$\xrightarrow{\frac{1}{2}\quad\frac{1}{2}} \qquad\qquad \xleftarrow{\frac{1}{2}}\xrightarrow{\frac{1}{2}}$$

0——1——2——3——4——5——6——7——8——9——10

4.2   Random Walk

We let x_n be the probability that A will win everything if he has n coins. (Geometrically, he is at point n.) Then we have

$$x_0 = 0 \qquad x_{10} = 1 \tag{4.23}$$

For $0 < n < 10$, a simple tree diagram shows that

$$x_n = \tfrac{1}{2}x_{n-1} + \tfrac{1}{2}x_{n+1} \qquad (0 < n < 10) \tag{4.24}$$

This system of equations constitutes 9 equations in 9 unknowns. We now seek its solution. To find it, we use a trick. Add $-\tfrac{1}{2}x_n - \tfrac{1}{2}x_{n-1}$ to both sides of Equation 4.24, to obtain $\tfrac{1}{2}x_n - \tfrac{1}{2}x_{n-1} = \tfrac{1}{2}x_{n+1} - \tfrac{1}{2}x_n$. Thus

$$x_{n+1} - x_n = x_n - x_{n-1} \qquad (n = 1, 2, \ldots, 9) \tag{4.25}$$

This is the system $x_1 - x_0 = x_2 - x_1 = x_3 - x_2 = \cdots = x_{10} - x_9$. Setting each of these numbers equal to k, we obtain

$$x_1 = x_0 + k, \; x_2 = x_1 + k = x_0 + 2k, \; x_3 = x_2 + k = x_0 + 3k, \ldots, x_{10} = x_0 + 10k$$

Using Equations 4.23 in the equation $x_{10} = x_0 + 10k$, we have $1 = 0 + 10k$. Thus $k = \frac{1}{10}$, and $x_n = x_0 + nk = 0 + n/10 = n/10$. Finally, the probability that A wins is $x_3 = \frac{3}{10}$. It is also seen that the probability that B wins is $x_7 = \frac{7}{10}$. Thus the game terminates with probability 1.

All these probabilities are to be interpreted as limits, as in Example 12. Note, however, that an *explicit* solution of this problem as a sum of an infinite series is not easy. It is easily seen that this example generalizes to an initial fortune of a for A and b for B. The probabilities of winning everything are proportional to the fortunes of A and B: $a/a+b$ for A and $b/a+b$ for B.

14 Example

In the above example, suppose A has probability p of winning a game and that B has probability $q = 1-p$ of winning. Again suppose that A starts with 3 pennies and B with 7, that the winner collects 1 cent after any win, and that the game terminates when one person has nothing left. What are the respective chances of winning everything?

The analysis is very similar to Example 13. Setting x_n = probability A will win everything if he has n pennies, we have again

$$x_0 = 0 \qquad x_{10} = 1 \tag{4.23}$$

However, this time the equations connecting the various probabilities x_n are

$$x_n = px_{n+1} + qx_{n-1} \qquad 0 < n < 10 \tag{4.26}$$

To solve these equations, subtract px_n from each side of Equation 4.26 to obtain

$$x_n - px_n = p(x_{n+1} - x_n) + qx_{n-1}$$

$$qx_n = p(x_{n+1} - x_n) + qx_{n-1}$$

$$q(x_n - x_{n-1}) = p(x_{n+1} - x_n)$$

$$x_{n+1} - x_n = \frac{q}{p}(x_n - x_{n-1}) \qquad (0 < n < 10) \tag{4.27}$$

Now set $r = q/p$ and $x_1 - x_0 = a$. Then if we take $n = 1, 2, \ldots, k$ in Equation 4.27, we obtain

$$x_1 - x_0 = a$$

$$x_2 - x_1 = ra$$

$$x_3 - x_2 = r(x_2 - x_1) = r^2 a$$

$$\vdots$$

$$x_k - x_{k-1} = \cdots = r^{k-1} a$$

Adding these equations, we have

$$x_k - x_0 = a(1 + r + \cdots + r^{k-1}) = a\frac{1-r^k}{1-r} \tag{4.28}$$

Recalling that $x_0 = 0$, $x_{10} = 1$, we find

$$1 = a\frac{1-r^{10}}{1-r}$$

$$a = \frac{1-r}{1-r^{10}}$$

Substituting in Equation 4.28 and using $x_0 = 0$, we finally obtain

$$x_k = \frac{1-r^k}{1-r^{10}} \qquad \left(r = \frac{q}{p} = \frac{1-p}{p}\right) \tag{4.29}$$

In our problem $k = 3$, so we had $x_3 = (1-r^3)/(1-r^{10})$, where $r = q/p$. The special case $p = \frac{1}{2}$ leads to $r = 1$, and in this case we are covered by Example 13, with $x_k = k/10$. The numerical results of Table 4.3 illustrate the relative importance of the size of p in comparison to how far behind a player is. Intuitively, if $p > .5$, there is a tendency in a random walk to creep to the right, while if $p < .5$, the drift is to the left.

4.3 Probability x_3 (to 3 Decimal Places) of Winning All, if Player Is Behind 3 to 7 but Has Probability p of Winning on Each Game

p	.1	.2	.3	.4	.5	.6	.7	.8	.9
r	9	4	$\frac{7}{3}$	$\frac{3}{2}$	1	$\frac{2}{3}$	$\frac{3}{7}$	$\frac{1}{4}$	$\frac{1}{9}$
x_3	.000	.000	.002	.042	.300	.716	.921	.984	.999

The method of this example clearly generalizes if there are N coins in circulation. We then have

$$x_k = \frac{1-r^k}{1-r^N} \qquad r = \frac{q}{p} \tag{4.30}$$

$$x_k = \frac{k}{N} \qquad (\text{if } p = \tfrac{1}{2}) \tag{4.31}$$

If we let $N \to \infty$ (with k fixed), we are in the position of playing an infinitely rich opponent. There are 3 cases: If $p < \frac{1}{2}$, then $r > 1$. Dividing numerator and denominator in Equation 4.30 by r^N, we have

$$x_k = \frac{(1/r^N) - (1/r^{N-k})}{(1/r^N) - 1}$$

Thus $x_k \to 0$ as $N \to \infty$. If $p = \frac{1}{2}$, we use Equation 4.31 and again $x_k \to 0$. But if $p > \frac{1}{2}$, then $r < 1$ and $r^N \to 0$. Thus $x_k \to 1 - r^k$. If we take complements, we have $1 - x_k = y_k = r^k =$ the probability of a player eventually losing his fortune. We summarize: *Suppose in a random walk a player moves one unit right with probability p, and one unit left with probability $q = 1-p$. Suppose he starts at $k > 0$. Then the probability that he will ultimately pass through $n = 0$ is 1 if $p \leqslant \frac{1}{2}$ but is $(q/p)^k$ if $p > \frac{1}{2}$.*[1]

We conclude this section by theoretically tying up the idea of the conditional probability $p(A|B)$ with the experimental idea.

15 Theorem

Let A and B be events with $p(B) > 0$. Let the experiment S be repeated indefinitely. Then the probability that A occurs when B first occurs is $p(A|B)$.

For a proof, let us restrict the number of experiments to at most n. Let T_n be the event (of S^n) that B occurs, and that A occurs also when B first occurs. Set $a = p(A \cap B)$, $c = p(\bar{B})$ and set $x_n = p(T_n)$. The tree diagram

$$x_n = a + cx_{n-1}$$

4.4 Tree Diagram for T_n

of Fig. 4.4 illustrates the two mutually exclusive ways in which T_n can occur, depending on what happens on the first experiment. Thus

$$x_n = a + cx_{n-1} \tag{4.32}$$

Since increasing n cannot decrease x_n, we have $x_n \to x$ by Theorem 10. Hence, taking limits in Equation 4.32, we have

$$x = a + cx$$

$$x = \frac{a}{1-c} = \frac{p(A \cap B)}{1 - p(\bar{B})} = \frac{p(A \cap B)}{p(B)} = p(A|B)$$

[1] The probability of eventually passing through $n = 0$, starting at $n = k$ should be interpreted as the limit of $P_{k,T}$, the probability of passing through $n = 0$ in T or less steps, as $T \to \infty$. The above derivation did not give this. Thus this derivation must strictly be regarded as heuristic, although it can be patched up.

Since x is simply "probability that A occurs when B first occurs," we have the result.

One way of looking at this theorem is to imagine an experiment S and two *mutually exclusive* events A and B. We then have a race: If the experiment is run repeatedly, which event will occur first, A or B? To find the probability that A wins, we find the probability that A occurs when $A \cup B$ first occurs. Thus, by Theorem 15, the probability that A occurs first is $p(A|A \cup B) = p(A)/(p(A)+p(B))$. Similarly, the probability that B occurs first is $p(B)/(p(A)+p(B))$. For example, suppose we toss 2 dice repeatedly. What is the probability that the sum 9 occurs before the sum 7? Here $p\,(\text{sum } 9) = \frac{4}{36}$, $p\,(\text{sum } 7) = \frac{6}{36}$. Hence $p\,(\text{sum } 9 \text{ before sum } 7) = \frac{4}{36}/(\frac{4}{36}+\frac{6}{36}) = \frac{4}{10} = \frac{2}{5}$.

EXERCISES

1. Suppose that players A and B have probabilities p and p', respectively. of hitting a target. They alternate shooting at the target, with A going first. What is the probability that A hits the target first? That B hits the target first?

2. In Exercise 1 prove that player A has a better chance than B of first hitting the target if and only if $p > p'/(1+p')$. Similarly, prove that A and B have equal chances if and only if $p = p'/(1+p')$.

3. Suppose in Exercise 1 that A has two chances at the target to every one that B has. Suppose $p = \frac{1}{3}$ and $p' = \frac{1}{2}$. Who has the better chance of first hitting the target?

4. Charlie and Doug play the following game. One of the players tosses a die. If it lands 5 or 6, he wins; if it lands 1, he loses; and otherwise, it becomes the other player's turn. What are Charlie's chances of winning if he starts the game? What are his chances of winning if the players toss a (fair) coin to decide who goes first?

5. Ed has \$3.00 and he intends to gamble incessantly until he wins \$5.00 or goes broke. Suppose that the probability of his winning a game is $p = .51$. Find the probability that Ed wins \$5.00 if he bets \$1.00 per game, and also if he bets \$.10 per game. Find these probabilities for $p = .50$ and for $p = .49$. (*Note:* Equation 4.10 and Appendixes A and B may be helpful.)

6. In Exercise 5, with $p = .51$, suppose Ed decides to play indefinitely or until he goes broke. What is the probability that he will go broke if he starts with \$3.00 and bets \$1.00 per game? What if he bets \$.10 per game?

7. Fred is tossing coins and is keeping a count of heads and tails. He finds that he has 100 more heads than tails. He claims that there is an excellent chance that heads will always be in the lead even if he continues forever. Discuss this claim.

8. The game of craps is played according to the following rules. Two dice are tossed and the sum S is observed. If $S = 2$, 3, or 12 the player loses. If $S = 7$ or 11, the player wins. But if $S = 4, 5, 6, 8, 9$, or 10 (called the players' "point") the player continues to toss the dice until his point is tossed again or until a 7 is tossed. If his point is tossed again, he wins; but if a 7 is tossed, he loses. Find the probability of winning.

9. Following Equation 4.22, it was stated that it seemed "clear" that x_n and y_n do not decrease with n. Explain that statement.

10. In Fig. 4.5 a particle starts at vertex A and moves along 1 of the 3 paths leading from A, each with probability $\frac{1}{3}$. Similar random movements occur at B and at C when the particle is there. The walk stops at X, Y, or Z. What is the probability the particle ends up at X? At Y? At Z?

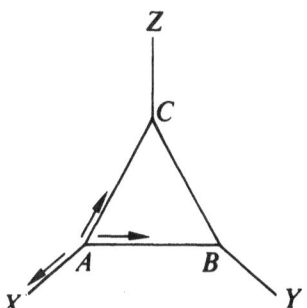

4.5   Random Walk About a Triangle

11. Three players, Xaviar, Yetta, and Zelda, play a fair game in which one of the players loses a penny to another, while the third does not gain or lose in this transaction. Suppose Xaviar, Yetta, and Zelda start with 1, 2, and 3 pennies, respectively, and that they keep playing until one person has no money left. What is the probability that Xaviar loses his money first? What is the probability that Yetta loses and that Xaviar has 3 or more pennies after Yetta loses? (*Hint:* Plot the fortunes of Xaviar and Yetta as a point in the xy plane and take a random walk.)

*3   COINCIDENCES

A dictionary[2] definition of "coincidence" is "a group of concurrent events or circumstances remarkable for lack of apparent causal connection." In this

2 *Webster's Seventh New Collegiate Dictionary*, G. & C. Merriam Company, Publishers, Springfield, Mass., 1965.

section we shall consider probabilistic explanations of certain types of co-incidences.

16 Example

Twenty people independently choose, at random, an integer between 1 and 100 inclusive. What is the probability that 2 of the people choose the same number?

Let p be the required probability. It is easier to compute $q = 1 - p$ = probability that all numbers are different. There are $_{100}P_{20}$ ways of choosing different numbers, and 100^{20} ways of choosing the numbers, all equally likely. Thus

$$q = \frac{_{100}P_{20}}{100^{20}} = \frac{100 \cdot 99 \cdot 98 \cdots (100-19)}{100^{20}} = \frac{100!}{100^{20}80!}$$

A computation using logarithms gives $q = .130$. Therefore, $p = 1 - q = .870$. Thus the "coincidence" of the same number appearing twice has a rather high probability of occurring. In a practical situation, unless they know how to choose numbers at random, people usually will choose special numbers. For example, they might tend to choose odd 2-digit numbers, making a coincidence even more likely.

We may write the expression for q as a product as follows:

$$q = \frac{(100-1) \cdot (100-2) \cdots (100-19)}{100^{19}}$$
$$= (1 - \tfrac{1}{100}) \cdot (1 - \tfrac{2}{100}) \cdots (1 - \tfrac{19}{100})$$

This expression for q may also be obtained directly by repeated application of the product rule (Section 3 of Chapter 3).

More generally, if r objects are chosen independently and with replacement from among n, the probability that no repetitions occur is

$$q = \frac{_nP_r}{n^r} = \left(1 - \frac{1}{n}\right)\left(1 - \frac{2}{n}\right) \cdots \left(1 - \frac{r-1}{n}\right)$$

Thus the probability that the same object is chosen twice is

$$p = 1 - \left(1 - \frac{1}{n}\right)\left(1 - \frac{2}{n}\right) \cdots \left(1 - \frac{r-1}{n}\right)$$

In analogy with Equation 4.7, we state without proof the following approximation for p.

17 Theorem

For any n and $r > 1$,

$$p = 1 - \left(1 - \frac{1}{n}\right)\left(1 - \frac{2}{n}\right) \cdots \left(1 - \frac{r-1}{n}\right) \approx 1 - e^{-r(r-1)/2n} \qquad (4.33)$$

In all cases $p > 1 - e^{-r(r-1)/2n}$.

We shall not concern ourselves with an estimate of the error in using Equation 4.33. Table 4.6 gives values of p and its estimate for various values of r and for $n = 100$. Further computations show that for $r = 13$, the probability of a match is greater than .5.

4.6 Probability p That When r Numbers Are Chosen at Random from n = 100 Numbers, 2 Will Be the Same

r	5	10	15	20	25	30
p	.097	.372	.669	.870	.962	.992
Estimate for p	.095	.362	.650	.850	.950	.987

It should be noted that the coincidence probability p is roughly dependent upon the *square* of r and inversely dependent on n. Thus, if n is increased, it is not necessary to increase r proportionately to achieve the same probability of a match. For example, when $n = 100$, we found that $r = 20$ yielded $p > .85$. If n is increased to 1,000, it is not necessary to increase r to 200 to yield $p > .85$. For if $n = 1,000$, we can find $p > .850$ by finding $e^{-r(r-1)/2,000} < .150$. Appendix A shows that this can be achieved if $r(r-1)/2,000 > 1.9$, because $e^{-1.90} = .1496$. Thus $r(r-1) > 3,800$. It suffices to take $r = 63$. Thus if 63 numbers are chosen at random from 1 to 1,000 inclusive, two of these numbers will be identical with probability larger than .85. Similarly, if we want a probability larger than .99, it suffices to take $r = 98$.

We can see this coincidence occurring in an everyday situation. If 30 people are gathered in a room, what is the probability that at least 2 of them have the same birthday? Making our usual prohibition against February 29, assuming all birthdates are equally likely and assuming independence of birthdates, we are in the situation discussed above with $n = 365$, $r = 30$. Thus $r(r-1)/2n = 30 \cdot 29/2 \cdot 365 = 1.19$. Using Appendix A, we have $p \approx 1 - e^{-1.19} = 1 - .304 = .696$. The chances are favorable that this coincidence occurs. It turns out, in fact, that for $r = 23$ there is a better than even chance that 2 people have the same birthday. For $r = 50$, using Equation 4.33, we find that $p > .975$.

Before considering another type of coincidence, it is convenient to introduce the following interesting generalization of Theorems 6 and 37 of Chapter 3.

18 Theorem. The Inclusion–Exclusion Principle

Let $A_1, \ldots, A_r$ be events in a sample space S. Let

$$s_1 = \sum_i p(A_i)$$

$$s_2 = \sum_{i<j} p(A_i \cap A_j)$$

$$s_3 = \sum_{i<j<k} p(A_i \cap A_j \cap A_k)$$

$$\vdots$$

$$s_r = p(A_1 \cap \cdots \cap A_r)$$

Then

$$p(A_1 \cup \cdots \cup A_r) = s_1 - s_2 + s_3 - \cdots \pm s_r \qquad (4.34)$$

The idea of the equation is that $s_1 = \sum p(A_i)$ includes all the probabilities of sample points in $A_1 \cup \cdots \cup A_r$. However, this sum may count some points twice. Thus we subtract $s_2 = \sum p(A_i \cap A_j)$, which includes all probabilities counted more than once in s_1. However, we may have subtracted too much, so we continue the process with s_3, etc.

To prove Equation 4.34 in general, we choose a sample point s in $A_1 \cup \cdots \cup A_r$ and find out how many times (positive, negative, or zero) it is counted in the right-hand side of the Equation 4.34. Suppose that the sample point s is in exactly k of the sets A_i ($k > 0$). Then the probability $p(s)$ is counted k times in s_1, $\binom{k}{2}$ times in s_2 (once for every pair of these k sets), $\binom{k}{3}$ times in s_3, etc. Thus $p(s)$ appears a total of

$$k - \binom{k}{2} + \binom{k}{3} - \cdots \pm \binom{k}{k}$$

times in the right-hand side of Equation 4.34. But this sum is exactly 1, because, by Equation 2.17,

$$1 - \binom{k}{1} + \binom{k}{2} - \cdots \pm \binom{k}{k} = 0 \qquad (k \geq 1)$$

Thus $p(s)$ appears exactly once in the right-hand side of Equation 4.34. This completes the proof.

The method of proof may be illustrated with the help of Fig. 4.7. Here $a, b, c, \ldots, k, l$ represent probabilities of the indicated points. Thus equation 4.34 becomes

$$a+b+c+d+e+f+g+h+i+j$$
$$= (a+b+c+d+e) + (d+e+f+g+h) + (c+e+g+h+i+j)$$
$$- [(d+e) + (c+e) + (e+g+h)]$$
$$+ e$$

It is seen that the term e corresponds to the case $k=3$ and it occurs exactly $3-3+1=1$ time. Similarly, g occurs $2-1=1$ time. This corresponds to $k=2$.

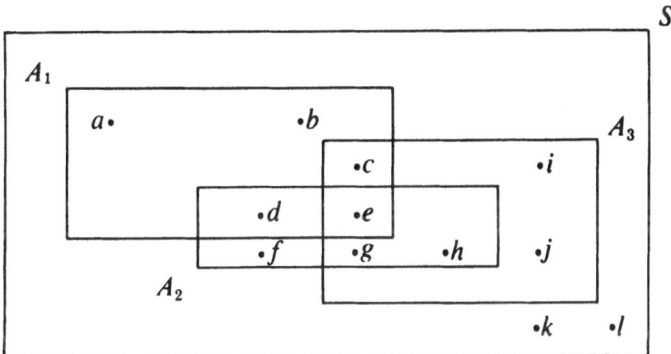

4.7 Inclusion–Exclusion Principle

Equation 4.34 may also be used for counting. If S contains k elements and if a uniform sample space is taken, then we have $n(A)=p(A)k$. Thus, multiplying Equation 4.34 by k, we obtain

$$n(A_1 \cup \cdots \cup A_r) = n_1 - n_2 + n_3 - \cdots \pm n_r \qquad (4.35)$$

where

$$n_1 = \sum_i n(A_i)$$

$$n_2 = \sum_{i<j} n(A_i \cap A_j), \qquad \text{etc.}$$

19 Example

How many 3-digit numbers are there which begin with the digit 5, end with the digit 7, or use 3 consecutive digits in some order?

We let S be the set of 3-digit numbers. Then define $B_5 = $ the set of numbers of S which begin with 5; $E_7 = $ the numbers of S which end with 7; Cons $= $ the set of numbers that contain 3 consecutive digits. Then a simple counting argument gives

$$n(B_5) = 10 \times 10 = 100$$
$$n(E_7) = 9 \times 10 = 90$$
$$n(\text{Cons}) = 8 \times 3 \times 2 \times 1 - 2 = 46$$

(A number of Cons is determined by the lowest digit appearing in it—0, 1,...,7—and then the locations of the digits. The two cases 012 and 021 are then eliminated.) Thus

$$n_1 = 100 + 90 + 46 = 236$$

Also,
$$n(B_5 \cap E_7) = 10$$
$$n(B_5 \cap \text{Cons}) = 3 \times 2 = 6$$
$$n(E_7 \cap \text{Cons}) = 3 \times 2 = 6$$

Thus
$$n_2 = 10 + 6 + 6 = 22$$

Also,
$$n_3 = n(B_5 \cap E_7 \cap \text{Cons}) = 1$$

Thus
$$n(B_5 \cup E_7 \cup \text{Cons}) = 236 - 22 + 1 = 215$$

Thus there are 215 such numbers.

We may use Theorem 18 to solve an interesting problem in coincidences which goes back to the eighteenth century. We first illustrate a special case.

20 Example

Ten balls (numbered 1 through 10) are placed at random into 10 boxes (numbered 1 through 10), 1 ball to a box. What is the probability that at least 1 of the balls occupies a box of its own number? (If ball i is in box i we call this a *match*.)

We let A_i be the event that ball numbered i is in box number i ($i = 1, 2, \ldots,$ 10). Then we wish to find $p(A_1 \cup A_2 \cup \cdots \cup A_n)$.

To find $p(A_1)$ we note that ball 1 has probability $\frac{1}{10}$ of landing in box 1. Thus $p(A_1) = \frac{1}{10}$. Similarly, $p(A_i) = \frac{1}{10}$. Thus

$$s_1 = p(A_1) + \cdots + p(A_{10}) = 10(\tfrac{1}{10}) = 1$$

To find $p(A_1 \cap A_2)$ we use the multiplication principle to obtain $p(A_1 \cap A_2) = p(A_1) \cdot p(A_2 | A_1) = (\frac{1}{10})(\frac{1}{9})$. Similarly, $p(A_i \cap A_j) = 1/10 \cdot 9$ for $i < j$. Since there are exactly $\binom{10}{2} = 10 \cdot 9/2 \cdot 1$ pairs i, j with $i < j$, we have

$$s_2 = \sum_{i<j} p(A_i \cap A_j) = \frac{1}{10 \cdot 9} \cdot \frac{10 \cdot 9}{2 \cdot 1} = \frac{1}{2!}$$

In the same way we find $p(A_i \cap A_j \cap A_k) = 1/10 \cdot 9 \cdot 8$ and that there are $\binom{10}{3} = 10 \cdot 9 \cdot 8/3 \cdot 2 \cdot 1$ triples $i < j < k$. Thus

$$s_3 = \sum_{i<j<k} p(A_i \cap A_j \cap A_k) = \frac{1}{10 \cdot 9 \cdot 8} \cdot \frac{10 \cdot 9 \cdot 8}{3 \cdot 2 \cdot 1} = \frac{1}{3!}$$

Continuing in this way, we find

$$s_k = \frac{1}{k!}$$

Finally, using Equation 4.34, we have

$$p(A_1 \cup \cdots \cup A_{10}) = \frac{1}{1!} - \frac{1}{2!} + \frac{1}{3!} - \cdots - \frac{1}{10!} = .6321$$

Thus at least one ball is in its own box with probability almost $\frac{2}{3}$.

It is easily seen that this example can be generalized as follows: *If n objects are rearranged at random, then with probability*

$$p_n = 1 - \frac{1}{2!} + \frac{1}{3!} - \cdots \pm \frac{1}{n!} \qquad (4.36)$$

at least one of the objects will occupy its original position.

A calculation of p_n from Equation 4.36 shows that p_n changes very little once n is large enough. For example, $p_n = .6321$ to 4 decimal places for all $n \geq 7$. (See Table 4.8)

4.8 *Values of p_n (to 4 Decimal Places)*

n	1	2	3	4	5	6	7 and up
p_n	1.0000	.5000	.6667	.6250	.6333	.6319	.6321

If we consider the *infinite series*

$$1 - \frac{1}{2!} + \frac{1}{3!} - \cdots$$

and compare with Equation 4.5 (taking $x = -1$):

$$e^{-1} = 1 - 1 + \frac{1}{2!} - \frac{1}{3!} + \cdots$$

we see that $p_n \to 1 - e^{-1}$ as $n \to \infty$:

$$p_n \approx 1 - \frac{1}{e} = .632120 \ldots \qquad (n \text{ large}) \qquad (4.37)$$

As noted above, for n as small as 7, 4-place accuracy is achieved by this approximation.

We can find p_n experimentally with cards. Shuffle a deck well. Then, from the top of the deck, count off the cards, one by one, while calling off the cards in some definite order, say the ace through king of spades, followed by ace through king of hearts, etc. Then the probability of a match is $p_{52} = .63212$. (The error in approximating p_{52} by $1 - 1/e$ is smaller than 10^{-69}. We may use Equation 4.37 unhesitatingly.) Many people consider that a correct call is a remarkable coincidence.

Similarly, if letters are addressed to 10,000 people and if these letters are stuffed into envelopes and mailed at random to these people, then with probability .63212, one of these people will receive a letter addressed to him.

Using Equation 4.36, we may find the probability that *exactly r* matches occur in random rearrangement of n numbers. We let $P_{[r]}$ denote the probability of exactly r matches. (Here n is given.) Clearly $r = 0$ is the complement of the event "at least one match." Thus $P_{[0]} = 1 - p_n = 1 - 1/1! + 1/2! - \cdots \pm 1/n! \approx 1/e$. Again, we set $A_i = $ probability that the ith number is a match. As before, $p(A_i) = 1/n$. The probability that the number i and only the number i is matched is, by the multiplication principle, $(1/n)(1 - p_{n-1})$. Since there are n numbers, the probability of exactly one match is $n(1/n)$ $(1 - p_{n-1}) = 1 - p_{n-1}$. Thus $P_{[1]} = 1 - 1/1! + \cdots \mp 1/(n-1)!$. Similarly, the probability that only i and j $(i < j)$ match is $[1/n(n-1)](1 - p_{n-2})$. Since there are exactly $\binom{n}{2} = n(n-1)/2!$ pairs i and j, we find that

$$P_{[2]} = \frac{n(n-1)}{2!} \frac{1}{n(n-1)} (1 - p_{n-2}) = \frac{1}{2!} \left(1 - \frac{1}{1!} + \frac{1}{2!} - \cdots \pm \frac{1}{(n-2)!} \right)$$

In a similar way, $P_{[r]}$ may be found. We summarize in the following theorem.

21   Theorem

If n numbers are rearranged at random, the probability $P_{[r]}$ of exactly r matches is given by the equations

$$P_{[0]} = 1 - \frac{1}{1!} + \frac{1}{2!} - \cdots \pm \frac{1}{n!} \approx \frac{1}{e} = .36788$$

$$P_{[1]} = \frac{1}{1!} \left(1 - \frac{1}{1!} + \frac{1}{2!} - \cdots \mp \frac{1}{(n-1)!} \right) \approx \frac{1}{1!} \cdot \frac{1}{e} = .36788$$

$$P_{[2]} = \frac{1}{2!} \left(1 - \frac{1}{1!} + \frac{1}{2!} - \cdots \pm \frac{1}{(n-2)!} \right) \approx \frac{1}{2!} \cdot \frac{1}{e} = .18394$$

$$\vdots$$

$$\text{(4.38)}$$

$$P_{[r]} = \frac{1}{r!} \left(1 - \frac{1}{1!} + \frac{1}{2!} - \cdots \pm \frac{1}{(n-r)!} \right) \approx \frac{1}{r!} \cdot \frac{1}{e}$$

$$\vdots$$

$$P_{[n-1]} = \frac{1}{(n-1)!} \left(1 - \frac{1}{1!} \right) = 0$$

$$P_{[n]} = \frac{1}{n!}$$

The approximations are valid when n is large.

EXERCISES

Wherever applicable, use the approximation formulas 4.33 or 4.37.

1. Each person in a class of 10 people is told to go home and choose 1 card from a well-shuffled deck of cards. Approximately what is the probability that 2 of these people will choose the same card?

2. How many people should be chosen at random from a telephone book to be sure, with probability greater than 95 percent, that 2 of these people will have the same birthday?

3. Twenty people each choose 3 different integers at random from 1 through 10 inclusive. Approximately what is the probability that 2 of these people choose the same set of integers?

4. Using Equation 4.35, find how many 4-digit numbers have different digits, and either have 1 as first digit, 2 as second digit, 3 as third digit, or 4 as fourth digit. (*Hint:* Take $S =$ set of 4-digit numbers with different digits.)

5. Answer Exercise 4 if the numbers need not have different digits. (*Hint:* In this case the digits are independent, and Equation 3.26 may be used.)

6. Three dice are tossed. What is the probability that a 1, or a 6, or three consecutive numbers turn up?

7. How many integers from 1 through 9,000 are divisible by 6, 10, or 45?

8. Two people, independently and at random, choose 3 different integers from 1 through 10. What is the probability that some number will be chosen by both people? [*Hint:* Let A_i be the event "i is chosen by both people." Find $p(A_1 \cup \cdots \cup A_{10})$ from Equation 4.4.]

9. As in Exercise 8, suppose two subsets of size r are chosen independently and at random from a population of size n. Show that the probability that the sets have nonempty intersection is

$$\frac{\binom{r}{1}^2}{\binom{n}{1}} - \frac{\binom{r}{2}^2}{\binom{n}{2}} + \frac{\binom{r}{3}^2}{\binom{n}{3}} - \cdots.$$

In particular, prove that the above expression is equal to 1 if $n/2 < r \leqslant n$.

10. By considering the complement event, show that the probability in

Exercise 9 is $1 - \left[\binom{n-r}{r} \Big/ \binom{n}{r} \right]$. Hence prove that

$$\frac{\binom{n-r}{r}}{\binom{n}{r}} = \frac{\binom{r}{0}^2}{\binom{n}{0}} - \frac{\binom{r}{1}^2}{\binom{n}{1}} + \frac{\binom{r}{2}^2}{\binom{n}{2}} - \cdots \pm \frac{\binom{r}{r}^2}{\binom{n}{r}} \tag{4.39}$$

11. A class of 30 takes seats at random in a classroom with a seating capacity of 30. The teacher then reads off her prearranged seating plan (also random).

    a. What is the probability that everybody has to move?
    b. What is the probability that exactly 3 people will not move?
    c. What is the probability that somebody does not move?
    d. What is the probability that 4 or more people will move?

*4 SYMMETRY

The notion that something is true "by symmetry" is often taken to be a rather vague, largely intuitive idea, whose purpose is to avoid a really mathematical argument. However, symmetry is a valid, rigorous, notion which appears in many parts of mathematics, and in this section we go into its applications to probability theory.

22 Example

If 5 dice are tossed, the probability that at least two 6's occur is .196. What is the probability that at least two 3's occur?

Clearly, the answer is also .196 "by symmetry." By this we mean that the roles of 3 and 6 are interchangeable in any consideration of this problem. Any derivation of the probability .196 can as well be done if the number 6 is replaced by 3.

To generalize this notion of symmetry, it is required to consider substitutions of 1 sample point for another. In the above dice problem we would substitute 6 for 3 and 3 for 6 in all sample points. This procedure is formalized by the notion of a *transformation*, which may be defined on any set.

23 Definition

A *transformation* of a set A is a method of assigning to every element x of A an element x' of A such that

    a. For every x in A, there is one and only one $y = x'$ corresponding to it.
    b. For every y in A, there is one and only one x in A such that $y = x'$.

When we have a particular transformation, we indicate it informally by an arrow: $x \rightarrow x'$. If we wish to name the transformation, we use functional notation and write $x' = f(x)$. In Example 22 it was convenient to consider the interchange of the outcomes 3 and 6. We thus have the transformation of Fig. 4.9.

4.9 Interchanging 3 and 6

1	2	3	4	5	6
↓	↓	↓	↓	↓	↓
1	2	6	4	5	3

In functional notation we may write

$$f(3) = 6$$
$$f(6) = 3$$
$$f(x) = x \qquad \text{for } x = 1, 2, 4, 5$$

Remark. A transformation of a set A is also called a *permutation* of A. This is seen to coincide with our previous notion (p. 54) if A is a finite set. In Fig. 4.9, for example, we may regard the interchange of 3 and 6 as the rearrangement 1 2 6 4 5 3, or as the ordered sample $(1, 2, 6, 4, 5, 3)$ without replacement from the set $\{1, 2, 3, 4, 5, 6\}$.

In the language of set theory, a transformation of a set A is a one-to-one mapping of the set A onto itself.

We can now define a symmetry.

24 Definition

Let S be a probability space and let $s' = f(s)$ be a transformation of S. We say that f is a *symmetry* of S if $p(s) = p(s')$ for all sample points s.

In brief, it is required that sample points which correspond to each other shall have equal probability.

If S is a *uniform* space, then clearly any transformation of S is a symmetry. But if S is not a uniform space, the symmetries are more limited. In all cases, however, our intuitive idea that *two sample points s and t play the same role*, is equivalent to the more precise idea that *some symmetry makes s correspond to t*.

If f is a symmetry of S, then f may be extended to a symmetry of $S \times S$ and in general to S^n. We merely define $f(x_1, \ldots, x_n) = (f(x_1), \ldots, f(x_n))$. For example, we may use the transformation of Fig. 4.9 and apply it to all sample points of Example 22. Thus we would have $(4, 6, 2, 6, 3) \to (4, 3, 2, 3, 6)$. If f is a symmetry of S, then subsets A of S can be made to correspond using f. Thus we define $f(A) =$ the set of all x', where $x \in A$. Briefly, we replace each x of A by $x' = f(x)$. We also write $A' = f(A)$.

25   Theorem

Let $s' = f(s)$ be a symmetry of the sample space S. Then if A is any event of S,

$$p(A') = p(A) \tag{4.40}$$

The proof is immediate. We have

$$p(A') = \sum_{s' \in A'} p(s') \qquad \text{[definition of } p(A')\text{]}$$

$$= \sum_{s \in A} p(s') \qquad \text{(definition of } A')$$

$$= \sum_{s \in A} p(s) \qquad \text{(definition of symmetry)}$$

$$= p(A) \qquad \text{[definition of } p(A)\text{]}$$

To see how this theorem is used, let us reconsider Example 22. If $A = \{1, 2, \ldots, 6\}$ is a uniform space, then the interchange $6 \to 3$, $3 \to 6$ is a symmetry.[3] This clearly defines a symmetry on $S = A^5$. If B is the event "at least two 6's," then B' is the event "at least two 3's." Thus $p(B) = p(B')$ by Equation 4.40. For the remainder of this section, we offer other illustrations of Theorem 23.

26   Example

Five dice are tossed. Show that the probability that the sum is 13 is the same as the probability that the sum is 22.

If $A = \{1, 2, 3, 4, 5, 6\}$ is the sample space of 1 die, we set up the symmetry

1	2	3	4	5	6
↓	↓	↓	↓	↓	↓
6	5	4	3	2	1

In general, $x' = f(x) = 7 - x$. This symmetry then extends to A^5. If $(x_1, x_2, x_3, x_4, x_5)$ is in the event "sum 13," we have $x_1 + x_2 + x_3 + x_4 + x_5 = 13$. Then $x_1' + \cdots + x_5' = (7 - x_1) + \cdots + (7 - x_5) = 35 - (x_1 + \cdots + x_5) = 22$. Conversely, if $x_1' + \cdots + x_5' = 22$, we obtain $x_1 + \cdots + x_5 = 13$. Thus

$$(\text{sum is 13})' = (\text{sum is 22})$$

By Equation 4.40 we have the result.

This symmetry is particularly easy to visualize. On the usual die, the opposite faces add up to 7: $(4, 3)$, $(5, 2)$, $(6, 1)$ are opposite faces. Thus the symmetry above has the effect of viewing 5 dice from "under the table." Since the "over plus under" sum is 7 for 1 die, it is 35 for 5, and hence the

---

3 In any such correspondence, it is understood that all elements x which are not mentioned stay fixed. Thus $x \to x$ if $x \neq 3$ or $x \neq 6$.

sum 13 (over the table) is just as likely as the sum 22 (under the table). In general, of course, any sum n has the same probability as the sum $35 - n$.

The transformation $x \rightarrow 7 - x$ also converts "high" into "low." Thus, for example, when 5 dice are tossed, the probability that the highest number is 5 is equal to the probability that the lowest number is 2. Similarly (Exercise 6, Section 1 of Chapter 1), if 5 dice are tossed and the numbers are arranged in increasing order, then the middle number is as likely to be 5 as 2. The reason in both cases is symmetry (i.e., Equation 4.40).

27 Example

If 3 cards are chosen from a deck (without replacement), what is the probability that n cards are black ($n = 0, 1, 2,$ and 3)? (This is Example 2.15, reconsidered.)

We let the probabilities be $p_n(n = 0, 1, 2,$ and 3). We note that there is an easy symmetry which interchanges black and red. (For example, we have the symmetry clubs $\leftrightarrow$ diamonds, spades $\leftrightarrow$ hearts, keeping the rank unchanged.) Under this symmetry, the event "no blacks" corresponds to "3 blacks." Similarly, "1 black" corresponds to "2 blacks." Thus $p_0 = p_3$, $p_1 = p_2$ by symmetry. Since $p_0 + p_1 + p_2 + p_3 = 1$, we have $2p_0 + 2p_1 = 1$. Thus, once p_0 is formed, p_1, p_2, p_3 are determined. Symmetry, however, does not give us the value of p_0.

28 Example

Ten coins are tossed. Show that the probability of 3 or more heads is the same as the probability of 7 or fewer heads.

Here we interchange heads and tails in the sample space $A = \{H, T\}$: $H \rightarrow T, T \rightarrow H$, and use this to define a symmetry on A^{10}. The event $B = $ "3 or more heads" corresponds to the event $B' = $ "3 or more tails" = "7 or less heads." Since $p(B) = p(B')$, we have the result.

It might be argued that we have a false result if the coin is unfair. But in this case the transformation $H \rightarrow T, T \rightarrow H$ would not be a symmetry, because it is required that $p(H) = p(H') = p(T)$. Thus the fairness of the coin was expressed by the statement that the above transformation was a symmetry. A look at Fig. 2.14 will convince the reader that the above symmetry translates into a geometric symmetry when properly graphed.

29 Example

Two different cards are successively chosen from a deck. What is the probability that the second card is higher in rank than the first? (Count the ace as the highest card.)

If the cards are (x, y), then the transformation $(x, y) \rightarrow (y, x)$ is a symmetry. Thus, by symmetry, it is equally likely that the second card is higher in rank than the first, as it is for it to be lower. Call this probability p. There is also a possibility that the cards have the same rank. This probability is $\frac{3}{51}$, regardless of the first card. Thus the probability that both have the same rank is $\frac{3}{51} = \frac{1}{17}$. Finally, the events considered are mutually exclusive and constitute all possibilities. Thus $1 = p + p + \frac{1}{17}$. Solving for p, we obtain $p = \frac{8}{17}$.

EXERCISES

1. If 3 dice are tossed, the probability that a 2 or 5 is tossed is $\frac{19}{27}$. What is the probability that a 4 or 6 is tossed? That a 1 or 2 is tossed? In each case state the symmetries involved.

2. Five dice are tossed. What is the probability that the sum of the faces is 17 or less?

3. Using the results of Table 2.4, find the probability that, when 3 dice are tossed, the number 5 occurs and, in addition, no even number is tossed. State your symmetry.

4. A die is tossed at most 3 times, or until a 1 or a 6 turns up. If a 1 turns up, Xaviar wins; if a 6 turns up, Yolanda wins; and if neither number occurs, Zelda wins. Using symmetry, show that Xaviar and Yolanda have equal chances of winning. Find the various probabilities of winning.

5. An urn contains 5 red, 5 green, and 7 white marbles. Three marbles are chosen at random (without replacement). Using symmetry as far as possible, find the probability that more reds are chosen than greens.

6. Prove: If f is a symmetry of S and if g is a symmetry of T, then the transformation h of $S \times T$, defined by $h(s, t) = (f(s), g(s))$, is a symmetry.

7. Three cards are chosen successively (without replacement) from a deck. What is the probability that they are chosen in increasing order (counting ace as high)?

8. An urn contains the integers 1 through 100. Three integers are successively chosen at random from the urn, without replacement. Find the probability they are chosen in increasing order.

9. In Exercise 8 find the probability if the ordered sample is done with replacement.

10. Generalize Exercises 8 and 9 if r integers are chosen from the integers 1 through n.

11. In Exercise 10 of Section 2, what can be said, using symmetry considerations, about the probabilities of ending at X, Y, or Z?

12.[4] In a well-shuffled deck, the ace of spades separates the remaining cards into two parts, those on top of it and those beneath it. For any integer n, prove that the probability that n cards are on top of this ace is equal to the probability that n cards are beneath it.

13. In Exercise 12 the 4 aces divide the deck into 5 sections. Using symmetry considerations, prove that for every integer n, the probabilities p_i = probability that the ith section has n cards ($i = 1, 2, \ldots, 5$) are equal. State the symmetry you are using.

---

4 Exercises 12 and 13 are discrete cases of a "principle of symmetry" discussed in F. Mosteller, *Fifty Challenging Problems in Probability*, Addison-Wesley Publishing Company, Inc., Reading, Mass., 1965, pp. 59–61.

CHAPTER 5 RANDOM VARIABLES

INTRODUCTION

Roughly speaking, a random variable is a number whose value is determined by the outcome of an experiment. It is a *variable* because it can be one of several numbers; it is *random* because the actual number depends on a probability experiment. Random variables have been part of probability theory since its beginnings, because they are invariably part of a gambling situation. For example, consider the following game in which 2 dice are tossed. The player wins $10 if both dice are 6 and he wins $1.00 if only 1 of the dice is 6. But he loses $1.00 if no 6 appears. Here the amount W of *winnings* in dollars is a random variable. The value of W can be 10, 1, or -1, depending on the outcome of the experiment.

We have already considered many random variables. A few examples are the number of heads among 10 tossed coins, the number of spades in a poker hand, and the relative frequency of an event A if an experiment is repeated independently for N times.

In this chapter we shall systematically investigate random variables and many important concepts related to them. It will be necessary to sum quantities extensively, so we shall start with a section on the Σ notation. We shall then use this notation more systematically and less informally than in the preceding sections.

1 SIGMA NOTATION

We have already used the symbol Σ as a general notation to designate a sum. We shall now go into more detail and study some of its properties.

1 Definition

Suppose that a and b are integers with $a \leq b$. Suppose that for each integer n between a and b inclusive, a number x_n is given. Then we define

$$\sum_{n=a}^{b} x_n = x_a + x_{a+1} + \cdots + x_b \qquad (5.1)$$

The left-hand side is read "sigma x sub n, $n = a$ to $n = b$." The numbers a and b are called, respectively, the lower and upper limits of the summation.

In case $b = a$, we naturally interpret Equation 5.1 to mean $\sum_{n=a}^{a} x_n = x_a$. Briefly, we substitute all values of n from a to b and then add the results. For example,

$$\sum_{n=2}^{4} x_n = x_2 + x_3 + x_4$$

$$\sum_{k=0}^{5} y_k = y_0 + y_1 + y_2 + y_3 + y_4 + y_5$$

$$\sum_{k=0}^{2} k^2 = 0^2 + 1^2 + 2^2 = 5$$

$$\sum_{i=1}^{4} (i+2) = (1+2) + (2+2) + (3+2) + (4+2) = 3+4+5+6 = 18$$

$$\sum_{j=4}^{7} 3 = 3+3+3+3 = 12$$

In the first two examples the specific values of x_n and y_k are not known. In the next two examples we have the formulas $x_k = k^2$ and $y_i = i+2$, respectively, so the sum can be evaluated. In the last example $x_j = 3$ for all values of j. Thus $x_4 + x_5 + x_6 + x_7 = 3+3+3+3$.

In a sum such as $\sum_{n=a}^{b} x_n$, the variable n is called a *dummy index*. The only role it plays is to be given values (from a to b) and replaced by these values. Any other letter can be used. Thus

$$\sum_{j=2}^{7} x_j = \sum_{k=2}^{7} x_k = \sum_{r=2}^{7} x_r$$

because each is shorthand for $x_2 + x_3 + \cdots + x_7$. A symbol such as $\sum_{n=1}^{n} x_n$ is sometimes used to designate $x_1 + \cdots + x_n$. This is not a good notation as it confuses the dummy variable n and the number of terms n. We rather write $\sum_{k=1}^{n} x_k$, in which the roles of k and n then become clear.

2 Example

Suppose $x_1 = 2$, $x_2 = 4$, $x_3 = 5$, and $x_4 = 9$. Compute $\Sigma_{i=1}^4 x_i$, $(\Sigma_{i=1}^4 x_i)^2$, $\Sigma_{i=1}^4 x_i^2$, and $\Sigma_{i=1}^4 ix_i$.

This is pure arithmetic:

$$\sum_{i=1}^4 x_i = x_1 + x_2 + x_3 + x_4 = 2 + 4 + 5 + 9 = 20$$

$$\left(\sum_{i=1}^4 x_i\right)^2 = (x_1 + x_2 + x_3 + x_4)^2 = 20^2 = 400$$

$$\sum_{i=1}^4 x_i^2 = x_1^2 + x_2^2 + x_3^2 + x_4^2 = 4 + 16 + 25 + 81 = 126$$

and

$$\sum_{i=1}^4 ix_i = x_1 + 2x_2 + 3x_3 + 4x_4 = 2 + 8 + 15 + 36 = 61$$

In what follows we shall usually state results summing from 1 to n rather than from a to b. In all cases there are analogous results for the limits a and b.

3 Theorem. Linearity of the Sum

$$\sum_{k=1}^n (a_k + b_k) = \sum_{k=1}^n a_k + \sum_{k=1}^n b_k \tag{5.2}$$

and

$$\sum_{k=1}^n ca_k = c \sum_{k=1}^n a_k \qquad \text{if } c \text{ is a fixed number} \tag{5.3}$$

For the proof, we compute. All summations are understood to have lower and upper limits 1 and n, respectively. We have

$$\sum (a_k + b_k) = (a_1 + b_1) + (a_2 + b_2) + \cdots + (a_n + b_n)$$
$$= (a_1 + a_2 + \cdots + a_n) + (b_1 + b_2 + \cdots + b_n)$$
$$= \sum a_k + \sum b_k$$

To prove Equation 5.3 we have

$$\sum ca_k = ca_1 + ca_2 + \cdots + ca_n$$
$$= c(a_1 + a_2 + \cdots + a_n) = c \sum a_k$$

Remark. Equation 5.2 may be generalized in the case where each summand is the sum of 3, or more, terms. For example,

$$\sum (a_k + b_k + c_k) = \sum a_k + \sum b_k + \sum c_k$$

This is an immediate consequence of Equation 5.2.

4 Theorem

For any constant c, we have

$$\sum_{k=1}^{n} c = nc \qquad (5.4)$$

For the proof we note that the left-hand side is the sum of n terms, each of which is c. Thus the sum is equal to nc.

Remark. The factor of c is always the number of terms. Since there are $b-a+1$ terms between a and b, inclusive, we have more generally

$$\sum_{k=a}^{b} c = c(b-a+1) \qquad (5.5)$$

The above results put into a compact notation algebraic manipulations involving adding the sums, factoring out a constant from a sum, and summing like terms. We can also add in the reverse order, using the following theorem.

5 Theorem

$$\sum_{k=1}^{n} a_k = \sum_{k=1}^{n} a_{n-k+1} \qquad (5.6)$$

For a proof, we note that

$$\sum_{k=1}^{n} a_{n-k+1} = a_{n-1+1} + a_{n-2+1} + \cdots + a_{n-n+1}$$

$$= a_n + a_{n-1} + \cdots + a_1$$

$$= a_1 + a_2 + \cdots + a_n = \sum_{k=1}^{n} a_k$$

The following theorem is useful for computing many sums. Some applications are given in the exercises.

6 Theorem. On Summing Differences

$$\sum_{k=1}^{n} (a_k - a_{k-1}) = a_n - a_0 \qquad (5.7)$$

For the proof, we write

$$\sum_{k=1}^{n} (a_k - a_{k-1}) = (a_1 - a_0) + (a_2 - a_1) + (a_3 - a_2) + \cdots + (a_n - a_{n-1})$$

In this sum all terms except a_0 and a_n cancel. Thus we have the result.

EXERCISES

1. Let x_n be given by the following table:

n	0	1	2	3	4	5
x_n	1.1	1.2	1.4	1.8	2.0	2.5

Compute:

a. $\displaystyle\sum_{n=0}^{5} x_n$ **b.** $\displaystyle\sum_{n=1}^{4} 10x_n$

c. $\displaystyle\sum_{n=1}^{5} (x_n - x_{n-1})$ **d.** $\displaystyle\sum_{n=0}^{5} (x_n - 1)^2$

e. $\displaystyle\sum_{n=1}^{5} \frac{x_n}{n}$ **f.** $\displaystyle\sum_{n=0}^{3} \left(\sum_{k=0}^{n} x_k\right)$

2. Express each of the following sums compactly by using the summation sign.

a $x_3 + x_4 + x_5 + \cdots + x_{24}$ **b.** $x_2 + x_4 + x_6 + \cdots + x_{18}$

c. $\frac{1}{2} + \frac{2}{3} + \frac{3}{4} + \cdots + \frac{9}{10}$ **d.** $1 - \frac{1}{2} + \frac{1}{3} - \frac{1}{4} + \cdots + \frac{1}{21}$

e. $1 + 2x + 3x^2 + 4x^3 + \cdots + nx^{n-1}$

3. Show that each of the following statements are true.

a. $\displaystyle\left(\sum_{k=1}^{n} x_k\right) + x_{n+1} = \sum_{k=1}^{n+1} x_k$

b. $\displaystyle\sum_{k=1}^{n} x_k + \sum_{k=n+1}^{m} x_k = \sum_{k=1}^{m} x_k$

c. $\displaystyle\sum_{k=1}^{n} x_{2k} + \sum_{k=1}^{n} x_{2k-1} = \sum_{k=1}^{2n} x_k$

d. $\displaystyle\sum_{j=1}^{n} \left(\sum_{k=1}^{j} x_k\right) = \sum_{k=1}^{n} (n-k+1)x_k$

4. If $x_1, \ldots, x_n$ are n numbers, express their average using the Σ notation.

5. Express $\Sigma_{i=1}^{n} (x_i - a)^2$ in terms of $\Sigma_{i=1}^{n} x_i^2$ and $\Sigma_{i=1}^{n} x_i$ by expanding and using the theorems proved in the text.

6. Express $\Sigma_{i=1}^{n} (x_i - y_i)^2$ in terms of 3 sums, as in Exercise 5.

7. Using Theorem 6 prove that $\Sigma_{k=1}^{n} [k^2 - (k-1)^2] = n^2$. Simplify the summand to prove $\Sigma_{k=1}^{n} (2k-1) = n^2$. Hence prove $1+2+\cdots+n = \Sigma_{k=1}^{n} k = n(n+1)/2$.

8. In analogy with Exercise 7, show that $\Sigma_{k=1}^{n} [k^3 - (k-1)^3] = n^3$. By simplifying and using the result proved in Exercise 7, show that

$$1^2 + 2^2 + \cdots + n^2 = \sum_{k=1}^{n} k^2 = \frac{n(n+1)(2n+1)}{6}$$

9. In analogy with Exercises 7 and 8, and using the results of those exercises, show that

$$1^3 + 2^3 + \cdots + n^3 = \sum_{k=1}^{n} k^3 = \frac{n^2(n+1)^2}{4}$$

10. Derive a formula for the sum

$$\sum_{k=a}^{b} (x_{k+1} - x_k)$$

11. Prove:

$$\sum_{k=a}^{b} x_k = \sum_{k=1}^{b-a+1} x_{k+a-1}$$

2 RANDOM VARIABLES AND THEIR DISTRIBUTIONS

The notion of a random variable was alluded to in the introduction to this chapter. We now give its formal definition.

7 Definition

Let S be a probability space. A *random variable* X is a real-valued function defined on S. Thus, given any sample point s of S, there is a uniquely determined value $X(s)$.

In theory, for finite sample spaces S, a random variable $X(s)$ can be quite arbitrarily given by a table of values. Thus Table 5.1 gives a sample space S together with a random variable on S.

5.1 A Random Variable

s	A	B	C	D	E
$p(s)$	.1	.2	.3	.1	.3
$X(s)$	2	3	4	2	3

However, random variables are usually defined in more natural ways, and the values are then found directly from the definition.

8 Example

Three coins are tossed. Construct the table for the number of heads that turn up.

We let $X =$ the number of heads. Clearly, the value of X depends upon the particular occurrence s. Thus $X = X(s)$ is a random variable. Its values are given in Table 5.2. As in many practical problems, however, the choice of a sample space is somewhat arbitrary. It would also be natural to choose S and X as in Table 5.3. Note that the random variables of Tables 5.2 and 5.3 are different, because they operate on different sample spaces.

5.2 $X(s) = Number\ of\ Heads$

s	HHH	HHT	HTH	HTT	THH	THT	TTH	TTT
$p(s)$	$\frac{1}{8}$	$\frac{1}{8}$	$\frac{1}{8}$	$\frac{1}{8}$	$\frac{1}{8}$	$\frac{1}{8}$	$\frac{1}{8}$	$\frac{1}{8}$
$X(s)$	3	2	2	1	2	1	1	0

5.3 $X(s) = Number\ of\ Heads$

s	3 heads	2 heads	1 head	0 heads
$p(s)$	$\frac{1}{8}$	$\frac{3}{8}$	$\frac{3}{8}$	$\frac{1}{8}$
$X(s)$	3	2	1	0

It is possible to introduce the idea of a *distribution* of a random variable from which we may consider certain aspects of random variables *without any reference to sample spaces*. It then turns out that the distribution for the random variables of Tables 5.2 and 5.3 are identical.

9 Definition

Let $X(s)$ be a random variable and let x be one of the possible values of $X(s)$. Let $A(x)$ be the set of sample points s such that $X(s) = x$. Then the function $p(x) = p(A(x))$ is called the *distribution* of X. The variable x is assumed to range over all possible values of $X(s)$.

The number $p(x)$ is simply the probability that $X(s) = x$. We sometimes write $p(x) = p(X = x)$. In practice, we simply consider the numbers x in some order $x_1, \ldots, x_n$ and give the distribution $p(x)$ in tabular form. It is important, however, that the x's include all the possible values of $X(s)$.

10 Example

Find the distribution of the random variable X of Table 5.2.

The values of $X(s)$ are seen to be 0, 1, 2, and 3. To find $p(1)$, for example, we find $A(1)$ – the set of s for which $X(s) = 1$. By observation, $A(1) = \{HTT, THT, TTH\}$. Thus $p(1) = p(A(1)) = \frac{3}{8}$. The other values for $p(x)$ are similarly found as in Table 5.4. It is seen that Table 5.4 also gives the distribution for the random variable of Table 5.3.

5.4 Distribution of the Number of Heads Among 3 Tossed Coins

x	0	1	2	3
$p(x)$	$\frac{1}{8}$	$\frac{3}{8}$	$\frac{3}{8}$	$\frac{1}{8}$

11 Theorem

Let $p(x)$ be a distribution of a random variable, where $x = x_1, \ldots, x_n$ are all the possibilities for x. Then

$$\sum_{i=1}^{n} p(x_i) = 1 \tag{5.8}$$

$$0 \leqslant p(x_i) \leqslant 1 \qquad (i = 1, \ldots, n) \tag{5.9}$$

To prove this theorem, suppose that X is the random variable and set $A(x_i) = $ the set of s such that $X(s) = x_i$. Then the events $A(x_i)$ are clearly mutually exclusive, and their union is S. (This latter statement is the meaning of the hypothesis that the x_i are all the possibilities for x.) Thus $\sum_{i=1}^{n} p(A(x_i)) = 1$, by Equation 3.7. Since $p(A(x_i)) = p(x_i)$, we have the result.

Equations 5.8 and 5.9 show that a distribution $p(x)$ may be regarded as a probability space on the set $x_1, \ldots, x_n$ of real numbers. This is an important step, because it brings us out of the apparatus of set theory into ordinary algebra. The letters $x_1, \ldots, x_n$ do not represent outcomes or elements of a sample space. They represent numbers! It is because of this that we are able to draw meaningful graphs of distribution and to construct tables of certain types of distributions. We shall consider some special distributions in Chapter 7.

Equation 5.8 is often abbreviated $\sum p(x_i) = 1$ or even $\sum p(x) = 1$. In all cases, when no upper and lower limits of summation are given, it is understood that the sum extends over all possible ranges of the variable.

We can obtain a good intuition about the distribution $p(x)$ of a random variable by graphing the equation $y = p(x)$. When the x's are integers, it is customary to plot a bar graph, in which the value $y = p(x_i)$ is plotted for all x between $x_i - \frac{1}{2}$ and $x_i + \frac{1}{2}$. This gives the illusion of a more substantial graph than would be obtained if the n points $(x_i, p(x_i))$ are plotted. Figure 5.5 illustrates 2 types of graphical representation for the distribution of Table 5.4. The scale along the y axis has been magnified for clarity. Note that the height over a number x is the probability that the random variable

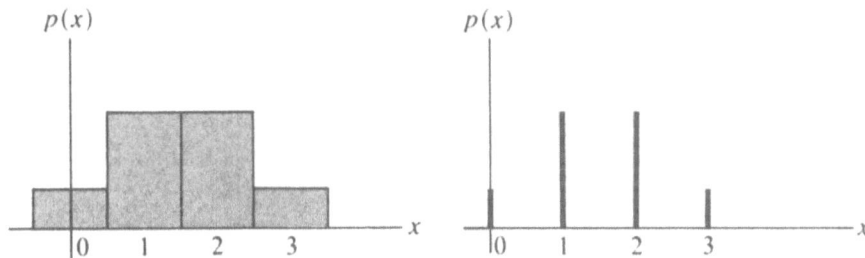

5.5 Graphs of a Distribution

attains the value x, so it is easy to see graphically where the more likely values of x are: These are the x values on which the heights are large. Bar graphs of distributions are often called *histograms*. The reader should refer to Fig. 2.14 for the graph of another distribution.

12 Example

A player tosses 2 dice. If two 6's occur, he wins $10.00. If only one 6 occurs, he wins $1.00. If no 6 occurs, he loses $1.00. Let $W =$ the winnings in dollars. Define the random variable W and give its distribution and its graph.

The natural probability space is A^2, where $A = \{1, 2, \ldots, 6\}$, the points on a die. (A is uniform.) The random variable W is defined by the formulas

$$
\begin{aligned}
W(x, y) &= 10 &&\text{if } x = 6 \text{ and } y = 6 \\
W(x, y) &= 1 &&\text{if } x = 6 \text{ and } y \neq 6 \\
W(x, y) &= 1 &&\text{if } x \neq 6 \text{ and } y = 6 \\
W(x, y) &= -1 &&\text{if } x \neq 6 \text{ and } y \neq 6
\end{aligned}
$$

The distribution $p(x)$ of W is defined for $x = 10$, 1, and -1. Its values are the probabilities of winning $10, $1, and $-$1, respectively. Thus we have the following table for $p(x)$:

x	-1	1	10
$p(x)$	$\frac{25}{36}$	$\frac{10}{36}$	$\frac{1}{36}$

Equivalently, we may write $p(-1) = \frac{25}{36}$, $p(1) = \frac{10}{36}$, $p(10) = \frac{1}{36}$. The graph (a histogram) is given in Fig. 5.6. The graph well illustrates a familiar aspect of gambling: a high yield with a low probability.

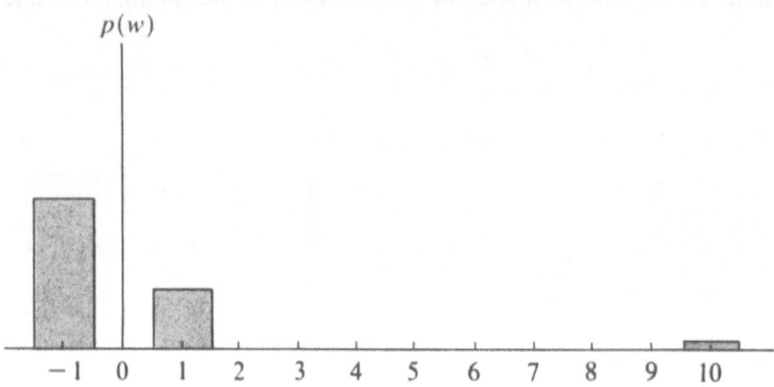

5.6 Graph of Distribution of Winnings

13 Example

Find the distributions of the random variables $X(s)$ and $Y(s)$ given by the following table:

s	a	b	c	d	e	f	g
$p(s)$	.1	.1	.1	.1	.2	.3	.1
$X(s)$	4	4	5	5	5	1	1
$Y(s)$	1	1	4	5	1	5	4

Here the possible values of x are $x = 1$, 4, or 5 in each case. If we let $p_1(x)$ be the distribution of X, and $p_2(x)$ the distribution of Y, we can compute $p_i(x)$ for each value of x. For $x = 1$, $p_1(1) = $ [probability that $X(s) = 1$] $= p(f) + p(g) = .3 + .1 = .4$. Similarly, $p_2(1) = $ [probability that $Y(s) = 1$] $= p(a) + p(b) + p(e) = .1 + .1 + .2 = .4$. In the same way we find that $p_1(4) = p_2(4) = .2$, and $p_1(5) = p_2(5) = .4$. Thus the random variables X and Y have identical distributions, *although X and Y are quite different random variables*. The common distribution $p(x)$ is given by the table

x	1	4	5
$p(x)$	.4	.2	.4

14 Example

Five cards are drawn, without replacement, from a deck. Find and graph the distribution for the number of black cards.

Here $x = 0, 1, \ldots, 5$ are possible. A counting argument gives

$$p(0) = p(5) = \binom{26}{5} \bigg/ \binom{52}{5} (= .025)$$

$$p(1) = p(4) = 26 \cdot \binom{26}{4} \bigg/ \binom{52}{5} (= .150)$$

$$p(2) = p(3) = \binom{26}{2}\binom{26}{3} \bigg/ \binom{52}{5} (= .325)$$

A bar graph of the distribution is given in Table 5.7. Note that the symmetry (interchanging red and black) shows up in the symmetry of the graph about the point $x = 2.5$.

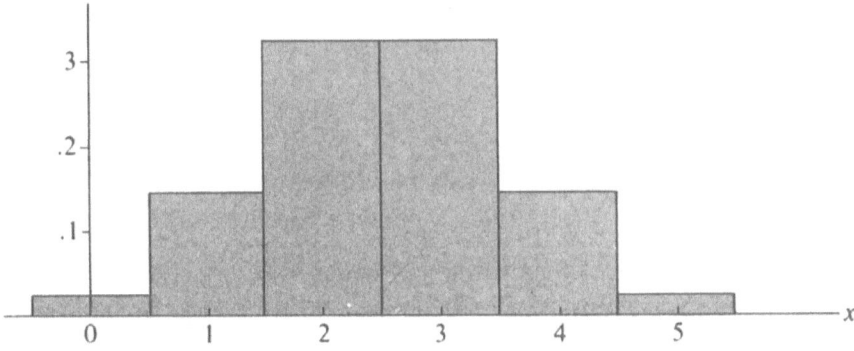

5.7 Number of Black Cards

15 Example

A die is tossed 13 times, or until a 6 turns up. Find and graph the distribution for the number of tosses.

Here $x = 1, 2, \ldots, 13$, and

$$p(x) = (\tfrac{5}{6})^{x-1}(\tfrac{1}{6}) \qquad x = 1, \ldots, 12$$

$$p(13) = (\tfrac{5}{6})^{12}$$

These values are tabulated in Table 1.6. The graph is given in Fig. 5.8. If the experiment is to continue indefinitely until a 6 occurs, the distribution would be defined for all positive integers x, and we would be out of the realm of finite sample spaces. We may regard the height at $x = 13$ in Fig. 5.8 as the infinite sum of the heights at $13, 14, \ldots$ in the unrestricted experiment.

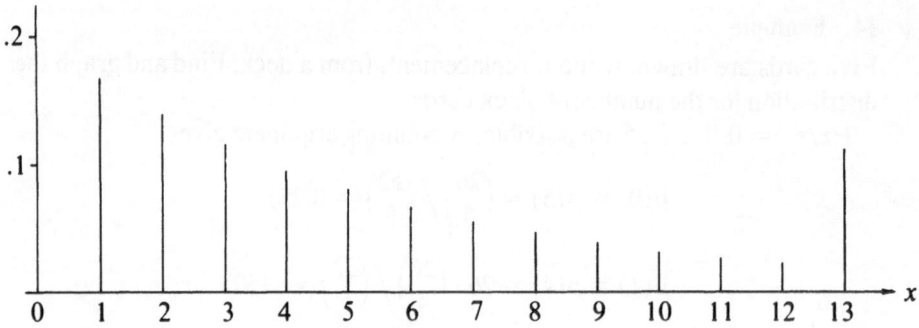

5.8   Number of Tosses Before a 6 Turns Up

EXERCISES

1. Let X be the random variable defined by the following table:

s	a	b	c	d	e
$p(s)$	.1	.4	.2	.2	.1
$X(s)$	1	2	3	2	1

Find the distribution of $X(s)$ and draw a bar graph of it.

2. If S is a probability space, is the probability $p(s)$ a random variable? If so, find the distribution of $p(s)$ if S is a uniform space and if S is the 5-point space of Exercise 1.

3. Two dice are tossed. Show that the sum S of the numbers tossed is a random variable, by giving an explicit formula for S. (Use the usual 36-point sample space.) Find the distribution of S and graph it.

4. As in Exercise 3, find the largest value M of the 2 numbers tossed, its distribution, and its graph. [Use the symbol max (x, y) to denote the maximum of x and y.]

5. Find the distribution for the number of heads among 5 tossed coins. Draw a graph of this distribution.

6. On a certain exam, the grades of the students were distributed as follows: Arthur 100 percent, Bernard 85, Carol 90, Doris 85, Eleanor 90, Frank 100, George 100, and Helene 100. Explain how the grade may be regarded as a random variable and find its distribution.

7. In playing the game of roulette, a player will win \$35.00 with probability $\frac{1}{38}$, but he will lose \$1 with probability $\frac{37}{38}$. Define and graph the distribution of his winnings W.

8. A person tosses a coin 6 times and computes the relative frequency f of heads. Explain why f is a random variable, and graph the distribution of f.

9. An 8-card deck consists of the 4 aces and the 4 kings. It is well shuffled. Compute and graph the distribution for the number of n of consecutive aces at the top of the deck. ($n = 0, 1, \ldots, 4$).

10. In Exercise 9 explain why the distribution for n is the same as the distribution for $m =$ the number of consecutive red cards at the top of the deck. Does this mean that the random variables m and n are equal? Explain.

3 EXPECTATION

If $X(s)$ is a random variable, it is sometimes convenient to find *one number* which is thought of as typical of the various numbers $X(s)$. The most commonly used, and the most useful for theoretical purposes, is the *expectation* of X, also called *expected value* of X, the *mean value* or *mean* of X, or the *average value* of X. It is directly related to the usual average of n numbers.

16 Example
Referring to Table 1.1, what was the average high number during the first 100 trials ? (The results are summarized below.)

Experimental Results: The High Number of 3 Dice

High number x	1	2	3	4	5	6
Frequency n	1	6	5	12	33	43

Although the high numbers were 1, 2, 3, 4, 5, and 6, it would be wrong to add these numbers and divide by 6. It is necessary also to consider the frequency of these numbers and to count each number as many times as it occurred. (Similarly, a student whose grades during a term were 100, 100, 100, and 50 does *not* add 100 and 50 and divide by 2 to obtain his average.) The average high number $\bar{x}$ can be computed as follows:

$$\bar{x} = \frac{1 \cdot 1 + 6 \cdot 2 + 5 \cdot 3 + 12 \cdot 4 + 33 \cdot 5 + 43 \cdot 6}{1 + 6 + 5 + 12 + 33 + 43}$$
$$= 4.99$$

We have used the formula

$$\bar{x} = \frac{\sum\limits_{i=1}^{6} n_i x_i}{\sum\limits_{i=1}^{6} n_i}$$

Here the numerator is the sum of all of the values x_i, each counted the proper number of times, while the denominator is the number of experiments. Each x_i occurs n_i times, so it contributes $n_i x_i$ to the sum. (The aforementioned student would find his average similarly: $\bar{x} = (3 \cdot 100 + 1 \cdot 50)/(3 + 1) = \frac{350}{4} = 87.5$.) The work in computing $\bar{x}$ may be done systematically using the format of Table 5.9, which is very convenient for use with a desk calculator.

5.9

x	n	$n \cdot x$
1	1	1
2	6	12
3	5	15
4	12	48
5	33	165
6	43	258
Sum	100	499

$$\bar{x} = \frac{\sum nx}{\sum n} = \frac{499}{100} = 4.99$$

This process is generalized as follows:

17 Definition

Suppose an experiment has outcomes given by a finite sample space $S = \{s_1, \ldots, s_k\}$ and that $X(s)$ is a random variable. Suppose that the experiment is performed N times and that the frequency of the outcome s_i is n_i (i.e., s_i occurred n_i times). Then the *average value* of X for this sequence of experiments is defined by the formula

$$\bar{x} = \frac{\sum n_i X(s_i)}{\sum n_i} = \frac{1}{N} \sum_{i=1}^{k} n_i X(s_i) \qquad (5.10)$$

We can write Equation 5.10 using *relative frequencies* f_i. We recall (Definition 2 of Chapter 1) that $f_i = n_i/N$. Thus using Equation 5.10 we have

$$\bar{x} = \frac{1}{N} \sum n_i X(s_i)$$

$$= \sum \frac{n_i}{N} X(s_i)$$

$$= \sum_{i=1}^{k} f_i X(s_i) \qquad (5.11)$$

In this formula there *is no division by N*. So to speak, the division has already been done in computing the relative frequencies f_i. We may illustrate Equation 5.11 taking the data of Example 16. The computation appears in Table 5.10.

5.10

x	n	f	fx
1	1	.01	.01
2	6	.06	.12
3	5	.05	.15
4	12	.12	.48
5	33	.33	1.65
6	43	.43	2.58
Sum	100	1.00	4.99

$$f = \frac{n}{N} = \frac{n}{100}$$

$$\bar{x} = \sum fx = 4.99$$

Equation 5.11 immediately suggests the definition of the average value of a random variable. This definition will be given only in terms of probabilities and will not involve experimental results. It is motivated by our original intention that, for a large number of experiments, $f_i \approx p_i = p(s_i)$. According to common usage, we use the term *expectation* of X, or *expected value* of X, rather than the equally valid terms *mean* of X, or *average value* of X. We also use the symbol $E(X)$ or $E(X(s))$ to denote the expected value of X.

18   Definition

Let $X = X(s)$ be a random variable on a finite sample space $S = \{s_1, \ldots, s_k\}$. The *expectation* of $X, E(X)$, is defined by the formula

$$E(X) = \sum_{i=1}^{k} p(s_i) X(s_i) \qquad (5.12)$$

This formula may also be written

$$E(X) = \sum p(s) X(s) \qquad (5.13)$$

or

$$E(X) = \sum p_i X(s_i) \qquad (5.14)$$

By comparing Equation 5.12 with 5.11 and recalling the notion of probability as a limiting relative frequency, we may give the following experimental significance of $E(X)$: *If a large number of experiments is performed, then the average of the experimental values of $X(s)$ will very likely be very close to $E(X)$.*

The *probabilities* for the outcomes in Example 16 were computed in Example 6 of Chapter 2. We may thus use these results and Equation 5.13 to compute $E(X)$, where X is the high number of 3 dice. The computation appears in Table 5.11. Thus $E(X) = 4.96$. The previous computation $\bar{x} = 4.99$ is quite close to $E(X)$.

5.11

x	p	px
1	1/216	1/216
2	7/216	14/216
3	19/216	57/216
4	37/216	148/216
5	61/216	305/216
6	91/216	546/216
Sum	1	1,071/216

$$E(X) = \frac{1,071}{216} = 4.96$$

Remarks.   The average value $\bar{x}$ of data is an abstract mathematical concept. Thus, when we find that the average family has 2.2 children or the average annual income in a certain city is $5,400 per year, there is no implication that any family has 2.2 children or that anyone earns $5,400 per year. The value $\bar{x}$ is an attempt to reflect complicated data (a *distribution* of values) with a single number. As such, it must be used with caution. For example, it may be reassuring to learn that the average income of a family in the United States is well above the poverty level, but this in no way replaces the more meaningful *distribution* of income. If 99 people in a town earn $1,000 per year, while 1 person earns $101,000 per year, the average income in dollars for these 100 people is ($99 \cdot 1,000 + 1 \cdot 101,000)/(99+1) = 2,000$. On the other hand, if all 100 people earned $2,000, the average income is again $2,000. We cannot expect one figure (the average) to replace the actual distribution.

Similarly, the number $E(X)$ is also an abstract mathematical concept, which is often even further removed from reality, because it is based on probabilities that must be either assumed or approximated. Nevertheless, we shall find ample use for this notion in the sequel.

We can use Equation 5.11 to compute experimental average values. All that we do is replace the probability p_i by the relative frequency f_i. This was justified in Section 5 of Chapter 1. However, if data are given with frequencies n_i, it is often useful to use Equation 5.10 rather than to compute relative frequencies.

The data of Example 16 were given in terms of the distribution of the random value $X(s)$. For example, even if the underlying sample space had $6^3 = 216$ sample points, there were only 6 possible values of $X(s)$, so the use of the distribution of $X(s)$ simplified the problem. The following theorem shows that we may compute $E(X)$ using only the distribution of X.

19   Theorem

If $p(x)$ $(x = x_1, \ldots, x_n)$ is the distribution of the random variable X, then

$$E(X) = \sum xp(x) = \sum_{i=1}^{n} x_i p(x_i) \tag{5.15}$$

To prove this, we write out the expression for $E(X)$ and collect all terms with the same $X(s)$. Thus, recalling Definition 9,

$$E(X) = \sum_{s \in S} p(s)X(s)$$

$$= \sum_{X(s)=x_1} p(s)X(s) + \sum_{X(s)=x_2} p(s)X(s) + \cdots + \sum_{X(s)=x_n} p(s)X(s)$$

$$= \sum_{X(s)=x_1} p(s)x_1 + \cdots + \sum_{X(s)=x_n} p(s)x_n$$

$$= x_1 \sum_{X(s)=x_1} p(s) + \cdots + x_n \sum_{X(s)=x_n} p(s)$$

$$= x_1 p(x_1) + \cdots + x_n p(x_n) = \sum_{i=1}^{n} x_i p(x_i)$$

We illustrate this theorem with a simple example.

20 Example

Three coins are tossed. Find the expected number of heads.

The computation using the definition is as follows. $X(s)$ is the number of heads in the sample point s:

s	HHH	HHT	HTH	HTT	THH	THT	TTH	TTT
$p(s)$	$\frac{1}{8}$	$\frac{1}{8}$	$\frac{1}{8}$	$\frac{1}{8}$	$\frac{1}{8}$	$\frac{1}{8}$	$\frac{1}{8}$	$\frac{1}{8}$
$X(s)$	3	2	2	1	2	1	1	0
$p(s)X(s)$	$\frac{3}{8}$	$\frac{2}{8}$	$\frac{2}{8}$	$\frac{1}{8}$	$\frac{2}{8}$	$\frac{1}{8}$	$\frac{1}{8}$	0

$$\sum p(s)X(s) = \tfrac{12}{8} = 1.5$$

Using the distribution of $X(s)$, the computation is as follows:

x	0	1	2	3
$p(x)$	$\frac{1}{8}$	$\frac{3}{8}$	$\frac{3}{8}$	$\frac{1}{8}$
$xp(x)$	0	$\frac{3}{8}$	$\frac{6}{8}$	$\frac{3}{8}$

$$\sum xp(x) = \tfrac{12}{8} = 1.5$$

If, for example, 10 coins were involved, the definition of $E(X)$ would give a sum with 1,024 terms, but Theorem 19 reduces this to 11 terms. This is the power of "collecting like terms."

21 Example

In Example 12, what are the expected winnings?

The distribution on page 161 is repeated here, and the computation is given in the table

x	$p(n)$	$xp(x)$
-1	$\frac{25}{36}$	$-\frac{25}{36}$
1	$\frac{10}{30}$	$\frac{10}{36}$
10	$\frac{1}{36}$	$\frac{10}{36}$
		$-\frac{5}{36}$ $\quad = \sum xp(x)$

We have $E(W) = -\frac{5}{36} = -.139$. Thus the player of that game loses, on the average, about $\$.14$ per game.

What is the significance of this figure? For the gambling house that offers this game, the figure $-\$.139$ is very meaningful. For they are concerned with a large number of "experiments." They know that when 10,000 games are played, they will, in all likelihood, be $\$1,390$ ahead on this game (give or take a few hundred dollars). On the other hand, a person might decide to play this game only once or twice, so this long-range aspect will not appear so significant to him. Yet it seems reasonable to take the expected winnings as a fair measure of the value of this game to him. (This measure does not take into consideration the "value" of the gambling experience.) There are various theories that measure the value of a gambling situation to a person, but the most commonly accepted one, and the one we use, is to use the expected winnings as the measure of the value of the game.

22 Example

A die is tossed and the player wins in cents, the number appearing on the die. If a 6 occurs, then the player wins 6 cents and is also entitled to one other throw, winning the second number also. It costs 4 cents to play the game. Is it worth it?

x	1	2	3	4	5	6	7	8	9	10	11	12
$w = x - 4$	-3	-2	-1	0	1	2	3	4	5	6	7	8
$p(w)$	$\frac{1}{6}$	$\frac{1}{6}$	$\frac{1}{6}$	$\frac{1}{6}$	$\frac{1}{6}$	0	$\frac{1}{36}$	$\frac{1}{36}$	$\frac{1}{36}$	$\frac{1}{36}$	$\frac{1}{36}$	$\frac{1}{36}$
$wp(w)$	$-\frac{3}{6}$	$-\frac{2}{6}$	$-\frac{1}{6}$	0	$\frac{1}{6}$	0	$\frac{3}{36}$	$\frac{4}{36}$	$\frac{5}{36}$	$\frac{6}{36}$	$\frac{7}{36}$	$\frac{8}{36}$

$$E(W) = wp(w) = -\frac{5}{6} + \frac{33}{36} = \frac{3}{36} = \frac{1}{12}$$

The above tabulation shows that the expected value of the winnings W is $\frac{1}{12}$. The game is favorable. Here, x is the amount won, and $w = x - 4$ is the net winnings, after paying the entrance fee.

Although we shall not use the concept in what follows, it is worth mentioning another important number, called the *median*, which is associated with a random variable or its distribution. Roughly speaking, the median is the value x_{med} such that it is just as likely that the value $X(s)$ is above it as below it. In terms of frequencies, half the results will be below the median and half above it. We illustrate with 2 examples and then leave the formal definition to the reader.

23 Example

Find the median value of x for the distribution of the following table:

x	1	2	3	4	5	6
$p(x)$	.11	.21	.03	.46	.08	.11

To do this we find the *cumulative probability* $P(x) =$ probability that $X(s) \leqslant x$. [For example, $P(3) = p(1) + p(2) + p(3)$ in this case.] The computation is

x	1	2	3	4	5	6
$p(x)$	.11	.21	.03	.46	.08	.11
$P(x)$	.11	.32	.35	.81	.89	1.00

The median $x_{med} = 4$. This is where the cumulative distribution first exceeded .5.

24 Example

Find the median value of x for the distribution of the following table:

x	1.1	1.3	1.8	2.1	2.5
$p(x)$	.31	.14	.05	.15	.35
$P(x)$	.31	.45	.50	.65	1.00

In this distribution the cumulative distribution is exactly .50 at $x = 1.8$. This means that $X(s) \leqslant 1.8$ with probability .5 but also that $X(s) \geqslant 2.1$ with probability .5. In this case it is customary to split the difference and to take the average $(\frac{1}{2})(1.8 + 2.1) = 1.95$ as the median. Note, however, that if we had $p(1.8) = .06$ and $p(2.1) = .14$, we would have taken $x = 1.8$ as the median.

As these examples show, the median is completely insensitive to the extreme values of x. If, in Example 23, we changed $x = 6$ to 6,000 and $x = 1$ to $x = -1,000,000$, the median would have remained 4. It is this insensitivity to extremes that often dictates the use of the median. On page 169 (Remarks) we gave the example of 99 people earning $1,000 and 1 earning $101,000. The average was $2,000, and it was that high because of the one atypical income. The median is, of course, $1,000.

On the other hand, a gambling house, or an insurance company, is quite sensitive to extremes. Thus, if an insurance company occasionally will have to pay off a large sum, it makes quite a difference to that company whether it is $1,000,000 or $100,000,000. Therefore, we may be certain that the insurance company will use the expected value of their payoffs, rather than the median, when they compute their rates.

EXERCISES

1. The daily maximum temperature in April 1966 in New York City is given in the table below. Compute the average maximum temperature for April. Also find the median maximum temperature.

Date	1	2	3	4	5	6	7	8	9	10	11	12	13	14	15
Max. temp. (°F)	49	49	56	53	54	55	57	52	53	55	57	59	44	60	66

Date	16	17	18	19	20	21	22	23	24	25	26	27	28	29	30
Max. temp. (°F)	62	63	66	57	47	69	64	68	55	73	73	52	42	60	55

2. Five coins are tossed. What is the average number of heads that turn up?

3. A die is tossed and the number x turns up. What is the average value of x? What is the average value of x^2?

4. Verify Theorem 19 for the random variable X of Table 5.12.

5.12

s	a	b	c	d	e
$p(s)$	.1	.2	.2	.3	.2
$X(s)$	1	2	5	1	2

5. Verify Theorem 19 for the random variable Y of Table 5.13.

5.13

s	A	B	C	D	E	F	G
$p(s)$	.1	.1	.1	.2	.2	.2	.1
$Y(s)$	3	3	3	3	5	5	5

6. Is it possible for the median of a random variable X to equal the mean of X? Give an example.

7. Two (different) cards are chosen from standard deck of cards. What is the average number of aces that are chosen? What is the average number of diamonds that are chosen?

8. Two dice are tossed. What is the expected value of the sum of the numbers that turn up?

9. Two dice are tossed. What is the expected high number that turns up? What is the expected low number?

10. Two dice are tossed. What is the expected difference between the high and low number?

11. Five coins are tossed. On the average, how many heads turn up?

12. Prove: If $a \leqslant X(s) \leqslant b$ for all sample points s, then $a \leqslant E(X) \leqslant b$.

13. Prove: If $X(s) = c$ (a constant) for all s, then $E(X) = c$.

14. An experiment has probability p of succeeding. It is repeated (independently) 3 times. What is the expected number of successes?

15. A whole number is chosen at random between 1 and 10 inclusive. On the average, how many divisors has the number? (For example, the number 10 has 4 divisors: 1, 2, 5, and 10.)

16. A ball is placed into one of 4 cups (numbered 1, 2, 3, and 4). This is repeated, independently, with 2 other balls. Let X be the largest number of balls occurring in a cup. (Thus $X = 1, 2,$ or 3.) Find $E(X)$.

17. In Exercise 16 let Y be the number of the first cup with at least 1 ball in it. (Thus $Y = 3$ means that cups 1 and 2 are empty but cup 3 has something in it.) Find $E(Y)$.

4 ALGEBRA OF EXPECTATIONS

There are a few theorems concerning the expectation $E(X)$ which simplify computations and which are of great theoretical importance. Before proceeding with these results, it is necessary to define the sum and product of random variables.

25 Definition
Let $X = X(s)$ and $Y = Y(s)$ be random variables defined on a probability space S. The *sum* $X+Y$ and the *product* XY are random variables on S defined by the formulas

$$(X+Y)(s) = X(s) + Y(s) \tag{5.16}$$

$$(XY)s = X(s)Y(s) \tag{5.17}$$

If c is any constant, we define the product cX by the formula[1]

$$(cX)(s) = c \cdot X(s) \tag{5.18}$$

26 Example
Let $X(s)$, $Y(s)$ be random variables as in Table 5.14. Find the random variables $X+Y$, XY, and $10Y$. Find the mean of X, Y, $X+Y$, XY, and $10Y$.

5.14

s	a	b	c
$p(s)$	.3	.5	.2
$X(s)$	1	2	3
$Y(s)$	3	1	1

Equation 5.16 shows that the values of $X+Y$ can be found by simply adding the values of X to the corresponding values of Y. XY and $10Y$ can be found similarly. The values of $X+Y$, XY, and $10Y$ are given in Table 5.15. The means are computed in the table. We note that $E(X) + E(Y) = 1.9 + 1.6 = 3.5$. Also, $E(X+Y) = 3.5$. Thus $E(X+Y) = E(X) + E(Y)$. Similarly,

---

1 Any constant c may be regarded as a random variable whose only value is c. With this understanding, Equation 5.18 is a special case of Equation 5.17, because

$$(cX)(s) = c(s)X(s) = c \cdot X(s)$$

5.15

s	a	b	c	Mean
$p(s)$	.3	.5	.2	
$X(s)$	1	2	3	$(1)(.3)+(2)(.5)+(3)(.2) = 1.9 = E(X)$
$Y(s)$	3	1	1	$(3)(.3)+(1)(.5)+(1)(.2) = 1.6 = E(Y)$
$(X+Y)(s)$	4	3	4	$(4)(.3)+(3)(.5)+(4)(.2) = 3.5 = E(X+Y)$
$(XY)s$	3	2	3	$(3)(.3)+(2)(.5)+(3)(.2) = 2.5 = E(XY)$
$10Y$	30	10	10	$(30)(.3)+(10)(.5)+(10)(.2) = 16 = E(10Y)$

$E(10Y) = 16$, while $10E(Y) = (10)(1.6) = 16$. Thus $E(10Y) = 10E(Y)$.[2]
We now state and prove these results in general.

27 Theorem

Let X and Y be random variables on the probability space S, and let c be a constant. Then

$$E(X+Y) = E(X)+E(Y) \tag{5.19}$$

$$E(cX) = cE(X) \tag{5.20}$$

$$E(c) = c \tag{5.21}$$

Proof. We merely calculate the left-hand side and reduce it to the right-hand side. By Equations 5.12 and 5.16 we have

$$E(X+Y) = \sum_{i=1}^{k} p(s_i)(X+Y)(s_i)$$

$$= \sum_{i=1}^{k} p(s_i)[X(s_i)+Y(s_i)]$$

$$= \sum_{i=1}^{k} [p(s_i)X(s_i)+p(s_i)Y(s_i)]$$

$$= \sum_{i=1}^{k} p(s_i)X(s_i) + \sum_{i=1}^{k} p(s_i)Y(s_i)$$

$$= E(X)+E(Y)$$

2 Note, however, that $E(X)E(Y) = (1.9)(1.6) = 3.04$, while $E(XY) = 2.5$. $E(XY) = E(X)E(Y)$ not true in general. Theorem 46 gives a condition under which this equation will be true.

Similarly,

$$E(cX) = \sum_s p(s)(cX)(s)$$

$$= \sum_s p(s)c \cdot X(s)$$

$$= c \sum_s p(s)X(s)$$

$$= cE(X)$$

Finally,

$$E(c) = \sum_{s \in S} cp(s)$$

$$= c \sum_{s \in S} p(s)$$

$$= c \cdot 1 = c$$

This proves the theorem.

If a random variable is the sum of more than 2 random variables, we can easily extend Equation 5.19. Thus $E(X+Y+Z) = E(X+Y)+E(Z) = E(X)+E(Y)+E(Z)$. In general, we have

$$E\left(\sum_{i=1}^n X_i\right) = \sum_{i=1}^n E(X_i)$$

28 Example

The average height, in feet, of the students in a class is 5.5. What is the average height, in inches?

Method.   One would suppose that the answer is $5.5 \times 12 = 66$ (inches), and it would appear that there is no problem. The reason for this supposition is that our feeling for changing units is ordinarily so ingrained that we take its properties for granted.

In this problem let the (uniform) sample space S be the set of students, and let $X(s) = $ the height, in feet, of students. Let $Y(s) = $ the height, in inches, of students. Then, because of the usual relationship between feet and inches, $Y(s) = 12X(s)$ for all s, or

$$Y = 12X$$

Therefore, by Equation 5.20,

$$E(Y) = E(12X) = 12E(X)$$

Since we were given $E(X) = 5.5$, it follows that $E(Y) = 12E(X) = 66$, as noted above.

This example shows that the formula $E(cX) = c \cdot E(X)$ may be interpreted as follows: *If the values of a random variable X are measured in certain units* (i.e., feet, dollars, cubic inches, etc.), *then the number $E(X)$ may be regarded as having that unit.*

29 Example

Ten coins are tossed. What is the average number of heads that turn up?

The sample space may be regarded as consisting of 1,024 sample points. For each sample point s, we let $N(s) =$ number of heads of s. Thus, if $s =$ HHTHTTTHTH, $N(s) = 5$. We wish to compute $E(N)$. We let

$$N_1(s) = \begin{cases} 1 & \text{if } s \text{ starts with H} \\ 0 & \text{if } s \text{ starts with T} \end{cases}$$

Briefly, $N_1(s) =$ number of heads on the first toss. Similarly, we let $N_2(s) =$ the number of heads on the second toss, and in the same way, $N_3, N_4, \ldots,$ N_{10} are defined. In each case, $N_i = 1$ or 0. [For the above sample point $s =$ HHTHTTTHTH, $N_1(s) = 1$, $N_5(s) = 0$, $N_{10}(s) = 1$.] The following remarks permit the calculation of $E(N)$:

(a) $N(s) = N_1(s) + N_2(s) + \cdots + N_{10}(s) = \Sigma_{i=1}^{10} N_i(s)$.
(b) $E(N_i) = \frac{1}{2}$.

To prove (a) note that the sum $\Sigma_{i=1}^{10} N_i(s)$ is a sum of 1's and 0's. Furthermore, each 1 occurs when, and only when, a head appears. Thus $\Sigma_{i=1}^{10} N_i(s) =$ number of heads in $s = N(s)$.

To prove (b) we can calculate the distribution $p_i(x)$ of N_i. Since $N_i(s) = 1$ if, and only if, s has a head on the ith occurrence, we see that the probability that $N_i = 1$ is $\frac{1}{2}$. Similarly, $N_i = 0$ with probability $\frac{1}{2}$. Thus the distribution $p_i(x)$ of N_i has the table

x	0	1
$p_i(x)$	$\frac{1}{2}$	$\frac{1}{2}$

Thus $E(N_i) = \Sigma\, x p_i(x) = 0 \cdot \frac{1}{2} + 1 \cdot \frac{1}{2} = \frac{1}{2}$. This proves (b). Finally we have, using (a) and (b),

$$E(N) = E\left(\sum_{i=1}^{10} N_i\right) = \sum_{i=1}^{10} E(N_i) = \sum_{i=1}^{10} \tfrac{1}{2} = 5$$

This example used an interesting idea. The number of heads was regarded as a sum of 0's and 1's, with summand 1 for each head. This idea will be used again in what follows.

30 Example

Five (different) cards are chosen at random from a standard deck. On the average, how many hearts are chosen?

Method. Let $N = N(s)$ be the number of hearts in the sample point s. As in Example 27, let $N_1(s) = 0$ if the first card is not a heart, and $N_1(s) = 1$ if the first card is a heart. If $N_2, \ldots, N_5$ are similarly defined, we have

$$N = N_1 + \cdots + N_5$$

$$E(N) = E(N_1) + \cdots + E(N_5)$$

But $E(N_1) = 0 \cdot \frac{3}{4} + 1 \cdot \frac{1}{4} = \frac{1}{4}$, because $N_1 = 1$ with probability $\frac{1}{4}$, and $N_1 = 0$ with probability $\frac{3}{4}$. Similarly, $E(N_i) = \frac{1}{4}$. Thus

$$E(N) = \frac{1}{4} + \frac{1}{4} + \cdots + \frac{1}{4} = \frac{5}{4}$$

and an average of $1\frac{1}{4}$ hearts are drawn. Note that a direct computation of $E(N)$, using Definition 18 or Theorem 19, is cumbersome.

EXERCISES

1. A box contains 50 black balls and 50 white balls. Thirty balls are selected (without replacement). What is the average number of black balls chosen?

2. In Exercise 1 suppose the 30 balls were chosen with replacement. What would the average number of chosen black balls be?

3a. A die is tossed. Let N be the number that turns up. Find $E(N)$.
b. Suppose 2 dice are tossed. Let S be the sum of the numbers turning up. Find $E(S)$. (*Hint:* Let $N_1 =$ the number on the first die and $N_2 =$ the number on the second die. Then $S = N_1 + N_2$.)
c. Suppose 10 dice are tossed. Let $S =$ the sum of all the numbers. Find $E(S)$.

4. Each time a certain computing machine adds 2 numbers, its probability of making an error is .0001. During the day this machine performs 100,000 additions. What is the expected number of errors?

5. Ten dice are tossed. What is the expected number of dice that turns up numbering 3 or higher?

6. One hundred cards are numbered from 1 through 100. Two of these cards are chosen (without replacement). What is the expected value of the sum?

7. In Exercise 6 suppose that x is the expected value of the high card chosen and y is the expected value of the low card. Evaluate $x + y$. (*Hint:* It is not necessary to evaluate x or y. This is a more difficult problem.)

8. Let A be a subset of a sample space S. Let $X(s) = 1$ for $s \in A$ and let $X(s) = 0$ for $s \in \bar{A}$. Prove $E(X) = p(A)$.

9. Suppose $E(X) = 1$, $E(X^2) = 2$. Find $E(X-1)$, $E(X-1)^2$, and $E(2X+1)^2$.

10. Suppose $E(X) = m$. Prove that $E(X-m)^2 = E(X^2) - m^2$.

11. Ten balls, numbered 1 through 10 are placed at random into 10 boxes similarly numbered. A match occurs if a ball is in a box with the same number. What is the expected number of matches? If the balls are placed at random into different boxes, what is the expected number of matches? (*Hint:* Let $M_1 = 1$ if ball 1 is in box 1, $M_1 = 0$ otherwise. Similarly define $M_2, \ldots, M_{10}$.)

5   CONDITIONAL EXPECTATIONS

Conditional expectations play a role in the calculation of expectations very similar to the role of conditional probabilities in the calculation of probabilities. We now define this notion and give some of its applications.

31   Definition
Let X be a random variable on the probability space S, and let A be an event of S with $p(A) > 0$. The *conditional expectation of X, given A* [written $E(X|A)$] is defined by the formula

$$E(X|A) = \sum_{s \in A} p(s|A)X(s) \tag{5.22}$$

Thus the value $E(X|A)$ may be regarded as "the expected value of X on the assumption that A has occurred." In order to relate $E(X|A)$ with an experimental situation, imagine running an experiment several times and computing the average value $\bar{x}$ of $X(s)$ *only for those s in A*. The relative frequency of an occurrence s (in A) will be $f(s|A)$. (See Definition 14 of Chapter 2.) By Equation 5.11 the resulting average $\bar{x}$ will be given by

$$\bar{x} = \sum_{s \in A} f(s|A)X(s)$$

Finally, if a large number of experiments are run, $f(s|A)$ will be near $p(s|A)$, so $\bar{x}$ will be near $E(X|A)$. Thus: *If a large number of experiments is performed and only the values $s \in A$ are counted, then the average of the experimental values of $X(s)$ will very likely be very close to $E(X|A)$.*

Conditional expectations may be put to use with the help of the following theorem, similar to the product rule for probabilities (Theorem 22 of Chapter 2) and the method of tree diagrams.

32   Theorem
Let $A_1, \ldots, A_n$ partition a probability space S, and assume that $p(A_i) > 0$. Let X be a random variable on S. Then

$$E(X) = p(A_1)E(X|A_1) + \cdots + p(A_n)E(X|A_n)$$

$$= \sum_{i=1}^{n} p(A_i)E(X|A_i) \tag{5.23}$$

Proof. We compute the right-hand side: For each i,

$$p(A_i)E(X|A_i) = p(A_i) \sum_{s \in A_i} p(s|A_i)X(s)$$

$$= \sum_{s \in A_i} p(A_i)p(s|A_i)X(s)$$

$$= \sum_{s \in A_i} p(s)X(s)$$

because $p(A_i)p(s|A_i) = p(s)$, by the multiplication rule. Thus

$$\sum p(A_i)E(X|A_i) = \sum_{s \in A_1} p(s)X(s) + \cdots + \sum_{s \in A_n} p(s)X(s)$$

$$= \sum_{s} p(s)X(s) = E(X)$$

This completes the proof.

Theorem 32 is useful when an experiment occurs in stages and tree diagrams can be used. The events A_i are thought of as the first stage. In Fig. 5.16 a random variable X is given whose expected value is 3, provided A_i has occurred, but whose expected value is 7 when A_2 has occurred. A_1 and A_2 occur with probabilities .9 and .1, respectively. (They are assumed mutually exclusive.) The expected value of X is computed to be 3.4.

$$\begin{array}{ll} {}_{.9}\;A_1\!\!-\!\!E(X|A_1) = 3 & (.9)\,(3) = 2.7 \\[6pt] {}_{.1}\;A_2\!\!-\!\!E(X|A_2) = 7 & (.1)\,(7) = \underline{\;.7\;} \\[4pt] & E(X) = \overline{3.4} \end{array}$$

5.16   Tree Diagram

33   Example

The average height of the boys in a class is 68 inches. The average height of the girls is 61 inches. Seventy-five percent of the class is male. What is the average height of a student in the class?

In the problem we regard relative frequencies as probabilities, invoking Section 5 of Chapter 1.[3] Then $H = H(s) =$ the height in inches of sample point s is a random variable on the sample space S of students. Let $B = $ the

---

3 For example, we can make the class into a uniform sample space. The set B of boys will then have probability $p(B) = .75$.

set of boys, G = the set of girls. We have

1. B and G partition S.
2. $p(B) = \frac{3}{4}, p(G) = \frac{1}{4}$.
3. $E(H|B) = 68; E(H|G) = 61$.

Thus, by Equation 5.22,

$$E(H) = p(B)E(H|B) + p(G)E(H|B) = (\tfrac{3}{4})(68) + (\tfrac{1}{4})(61) = 66.25$$

The computation is indicated in Fig. 5.17.

B——$E(H|B) = 68$ $(\tfrac{3}{4})(68)$

G——$E(H|G) = 61$ $(\tfrac{1}{4})(61)$
$$E(H) = (\tfrac{3}{4})(68) + (\tfrac{1}{4})(61) = 66\tfrac{1}{4}$$

5.17   Tree Diagram for Heights of Students

34   Example

A coin is tossed 5 times, or until a head turns up – whichever happens first. On the average, how many times is the coin tossed?

Method 1.   A direct computation is given in Fig. 5.18. The random variable N is the number of tosses for a given sample point s. Thus $N(\text{TTH}) = 3$. This computation shows that $E(N) = 1.9375$.

$H(N = 1)$ $p(1)N = (\tfrac{1}{2})(1) = \tfrac{1}{2}$

$T \longrightarrow H(N = 2)$ $p(2)N = (\tfrac{1}{4})(2) = \tfrac{1}{2}$

$T \longrightarrow H(N = 3)$ $p(3)N = (\tfrac{1}{8})(3) = \tfrac{3}{8}$

$T \longrightarrow H(N = 4)$ $p(4)N = (\tfrac{1}{16})(4) = \tfrac{1}{4}$

$T (N = 5)$ $p(5)N = (\tfrac{1}{16})(5) = \tfrac{5}{16}$

$$E(N) = \sum p_i N_i = \tfrac{1}{2} + \tfrac{1}{2} + \tfrac{3}{8} + \tfrac{1}{4} + \tfrac{5}{16} = 1\tfrac{15}{16}$$

5.18   Tree Diagram for Coin Tossing

Method 2.   Conditional expectations appear to be useful here, because the two cases (first toss H, or first toss T) can each be computed separately. To do this we must first consider a 4-toss game (in which at most 4 tosses are allowed). We then regard the given game in 2 stages: If the first toss is T, we play the 4-toss game, but if the first toss is H, we stop. We let $N_4 = N_4(s)$ be the number of tosses for 4-point games and take $N = N_5$. We then have (where s is any sample point in the 4-toss game)

$$N_5(H, s) = 1$$

$$N_5(T, s) = 1 + N_4(s)$$

Thus

$$E(N_5|H) = E(N_5(H, s)) = 1$$

and

$$E(N_5|T) = E(N_5(T, s)) = E(1 + N_4(s)) = 1 + E(N_4)$$

This is indicated in Fig. 5.19.

$$\imath \quad H{-}E(N_5|H) = 1$$

$$\imath \quad T{-}E(N_5|T) = 1 + E(N_4)$$

5.19 Another Tree Diagram for Coin Tossing

Thus, using Equation 5.23, we have

$$E(N_5) = (\tfrac{1}{2})(1) + (\tfrac{1}{2})[1 + E(N_4)]$$

$$= 1 + (\tfrac{1}{2})E(N_4) \qquad (5.24)$$

Equation 5.24 evaluates $E(N_5)$ in terms of $E(N_4)$. But in the same way, we can obtain

$$E(N_{k+1}) = 1 + (\tfrac{1}{2})E(N_k) \qquad (k = 1, 2, 3, 4) \qquad (5.25)$$

Thus

$$E(N_2) = 1 + (\tfrac{1}{2})E(N_1) = \tfrac{3}{2}$$

$$E(N_3) = 1 + (\tfrac{1}{2})E(N_2) = 1 + (\tfrac{1}{2})(\tfrac{3}{2}) = \tfrac{7}{4}$$

$$E(N_4) = 1 + (\tfrac{1}{2})(\tfrac{7}{4}) = \tfrac{15}{8}$$

and, finally,

$$E(N_5) = 1 + (\tfrac{1}{2})(\tfrac{15}{8}) = \tfrac{31}{16} = 1\tfrac{15}{16}$$

For this computation it is seen that regardless of how many tosses are allowed, the expected number of tosses before a head turns up is less than 2. In fact, if we continue in this way, we can show that

$$E(N_k) = 2 - 1/2^{k-1} \qquad (k = 1, 2, \ldots) \qquad (5.26)$$

It is natural to say that if an unlimited number of tosses were allowed (i.e., if there were no artifical stopping point), the expected number of tosses would be 2. This a correct statement, but it involves infinite sample spaces, as in Section 2 of Chapter 4.

35 Example

A die is tossed until a 6 turns up. What is the expected number N of tosses?

Method. This problem involves an infinite sample space, but we shall nevertheless illustrate the technique of Theorem 32 here. Two cases can occur: If the first toss is a 6, then $N = 1$. [$E(N|$ first toss is a 6$) = 1$.] If the first toss is not a 6, then $E(N|$ first toss is not a 6$) = 1 + E(N)$. For, if the first toss is not a 6, we are left in the same game, except that we have used up 1 toss. Thus we have the tree diagram of Fig. 5.20. Then, using Theorem 32, we have

$$E(N) = (\tfrac{1}{6})(1) + (\tfrac{5}{6})[1 + E(N)]$$
$$= 1 + (\tfrac{5}{6})E(N)$$

When this equation is solved for $E(N)$ we obtain $E(N) = 6$. This is the required expectation.

$$
\begin{array}{l}
{}_{\frac{1}{6}}\diagup^{6} \text{———} E(N|6) = 1 \\
\diagdown_{\frac{5}{6}} \text{not } 6 \text{———} E(N|\text{not } 6) = 1 + E(N)
\end{array}
$$

5.20   Tree Diagram for Dice

We finally give another direct illustration of Theorem 5.32.

36 Example

A whole number X is chosen at random between 1 and 4 inclusive. The number Y is then chosen at random between 1 and X. Find the expected value of Y.

Method.   If S is the appropriate sample space, it is clear that $X(s)$ and $Y(s)$ are well defined. Further, $E(Y|X = 1) = 1$. Similarly, $E(Y|X = 2) = (\tfrac{1}{2})(1) + (\tfrac{1}{2})(2) = \tfrac{3}{2}$, $E(Y|X = 3) = (\tfrac{1}{3})(1+2+3) = 2$, and $E(Y|X = 4) = (\tfrac{1}{4})(1+2+3+4) = \tfrac{5}{2}$. Thus we have Fig. 5.21, in which $E(Y)$ is computed to be $\tfrac{7}{4}$.

$$
\begin{array}{lll}
{}_{\frac{1}{4}}\; X = 1 \text{———} E(Y|X = 1) = 1 & \quad p(1)E = (\tfrac{1}{4})(1) \\
{}_{\frac{1}{4}}\; X = 2 \text{———} E(Y|X = 2) = \tfrac{3}{2} & \quad p(2)E = (\tfrac{1}{4})(\tfrac{3}{2}) \\
{}_{\frac{1}{4}}\; X = 3 \text{———} E(Y|X = 3) = 2 & \quad p(3)E = (\tfrac{1}{4})(2) \\
{}_{\frac{1}{4}}\; X = 4 \text{———} E(Y|X = 4) = \tfrac{5}{2} & \quad p(4)E = (\tfrac{1}{4})(\tfrac{5}{2})
\end{array}
$$

$$E(Y) = (\tfrac{1}{4})(1) + (\tfrac{1}{4})(\tfrac{3}{2}) + (\tfrac{1}{4})(2) + (\tfrac{1}{4})(\tfrac{5}{2}) = \tfrac{7}{4}$$

5.21   Tree Diagram

EXERCISES

1. Forty percent of the people in a certain community have an average annual income of $7,200. The remaining 60 percent have an average annual income of $9,000. What is the average annual income of the people in the community?

2. Two dice are tossed. If the sum is 8 or more, the player wins $1; otherwise, he wins nothing. If the sum is 10 or more, he is also entitled to another throw. On his second throw he wins, in dollars, the sum thrown on the second try. What are his expected winnings?

3. A die is tossed repeatedly until the sum of all the numbers tossed is 5 or more. (For example, if 1, 3, 4 are tossed in order, the play would stop after these 3 tosses, because $1+3+4$ is larger than 5.) On the average, how many times will the dice be tossed? (*Hint:* After the first die is tossed, the game is reduced to a similar game but with a smaller cutoff point. Thus consider the game where the play stops when the sum is 4, 3, 2, and 1.)

4. A coin is tossed until 2 consecutive heads occur. What is the average number of tosses required? (Ignore the fact that this game is potentially infinite in length.)

5. Using the result of Exercise 4, what is the average number of times it is required to toss a coin until 3 consecutive heads occur? Generalize, by finding the average number of tosses needed to toss k consecutive heads.

6a. If an integer N is chosen at random between 1 and n inclusive, prove that $E(N) = (n+1)/2$.
  b. Suppose that an integer N_1 is chosen at random between 1 and n inclusive, and then N_2 is chosen at random between 1 and N_1. Find $E(N_2)$.
  *c. Keep going! Choose N_3 at random between 1 and N_2. Find $E(N_3)$. Similarly, find a formula for $E(N_k)$, where N_k is the number chosen after k trials. [Your intuition should tell you that $E(N_k)$ is almost 1 when k is large.]

6   JOINT DISTRIBUTIONS

If X is a random variable on a probability space S, we have seen in Section 2 how many of the properties of X may be summarized by the distribution $p(x)$ of X. If X and Y are 2 random variables, *both operating on the same probability space*, we may analogously define the joint distribution of X and Y.

37 Definition

Let X and Y be random variables on a probability space S. Let x be one of the values of $X(s)$ and let y be one of the values of $Y(s)$. Let $A(x, y)$ be the set of sample points s such that $X(s) = x$ and $Y(s) = y$. Then the function $p(x, y) = p(A(x, y))$ is called the *joint distribution* of X and Y. The variables x and y are assumed to range over all possible values of $X(s)$ and $Y(s)$ respectively.

Remark. Using the notation of Definition 9, $A(x, y) = A(x) \cap B(y)$, where $A(x)$ is the set of s with $X(s) = x$ and $B(y)$ the set of s with $Y(s) = y$. $p(x, y)$ is simply the probability that $X(s) = x$ and $Y(s) = y$. We sometimes write $p(x, y) = p(X = x$ and $Y = y)$. The values of $p(x, y)$ can be conveniently recorded in a table, as Example 38 illustrates.

38 Example

Two coins are tossed. Let $X = 1$ if the first coin is head and $X = 0$ if the first coin is tails. Let $Y = $ number of heads tossed. Find the joint distribution $p(x, y)$ of X and Y.

Method. The values for x are 0 and 1; the values for y are 0, 1, and 2. Place the x values in a column and the y values in a row, as in Table 5.22. Then $p(x, y)$ is computed in all cases and entered in the row opposite x and column under y. For example, $p(0, 1) = $ probability the first coin is tail, and the number of heads (on 2 tosses) is 1. Thus $p(0, 1) = \frac{1}{4}$. In the same way, each of the 6 entries may be found.

5.22   Joint Distribution

x \ y	0	1	2	Total $p(x)$
0	$\frac{1}{4}$	$\frac{1}{4}$	0	$\frac{1}{2}$
1	0	$\frac{1}{4}$	$\frac{1}{4}$	$\frac{1}{2}$
Total $q(y)$	$\frac{1}{4}$	$\frac{1}{2}$	$\frac{1}{4}$	

Note that the sum of each row gives the distribution $p(x)$ of X, while the sum of each column gives the distribution $q(y)$ of Y. This will be proved in what follows, but it should be clear. For $p(0) = $ probability $(X = 0) = p(X = 0$ and $Y = 0) + p(X = 0$ and $Y = 1) + p(X = 0$ and $Y = 2)$. But this is precisely the sum of the first row, and similarly for all rows and columns.

39 Theorem

Let X and Y be random variables on a probability space S, and let $p(x, y)$ be their joint distribution. Then

$$p(x) = \sum_y p(x, y) \tag{5.27}$$

is the distribution of X, while

$$q(y) = \sum_x p(x, y) \tag{5.28}$$

is the distribution of Y.

Equation (5.27) expresses $p(x)$ as the sum of all the entries in the row corresponding to x, while (5.28) expresses $q(y)$ as the sum of all the entries in the columns corresponding to y. (See Table 5.22.) The distributions $p(x)$ and $q(y)$ obtained in this way from the joint distribution $p(x, y)$ are sometimes called *marginal distributions*.

Proof. Let $A(x) =$ the set of s where $X(s) = x$, and let $B(y) =$ the set of s where $Y(s) = y$. Then, if $y_1, y_2, \ldots, y_n$ are all possible values of y, we have

$$A(x) = \{A(x) \cap B(y_1)\} \cup \{A(x) \cap B(y_2)\} \cup \cdots \cup \{A(x) \cap B(y_n)\} \tag{5.29}$$

This equation merely states that $X(s) = x$ occurs with one of the possibilities $Y(s) = y_j$. But these possibilities are mutually exclusive. Therefore we can use Theorem 10 of Chapter 3 to obtain

$$p(x) = p(A(x)) = \sum_{j=1}^n p(A(x) \cap B(y_j))$$
$$= \sum_{j=1}^n p(x, y_j)$$
$$= \sum_y p(x, y)$$

This proves Equation 5.27. Equation 5.28 may be proved similarly.

40 Example

A whole number x is chosen at random between 1 and 4 inclusive. The number y is then chosen at random between 1 and x. Find the expected value of y.

Method. (Cf. Example 36.) We shall find the joint distribution $p(x, y)$, then the distribution $q(y)$, and then the expected value. Table 5.23 is computed separately along columns 1, 2, 3, and 4. For example, to find the third row, we must compute $p(3, y) =$ probability first number is 3 and the second is y. This is seen to be $(\frac{1}{4})(\frac{1}{3}) = \frac{1}{12}$ for $y = 1, 2, 3$ but is 0 for $y = 4$. The other rows are dealt with similarly. The distribution $q(y)$ is obtained by summing the columns. Finally, using Theorem 19, $E(Y) = \sum y q(y) = \frac{7}{4} = 1.75$.

5.23

x \ y	1	2	3	4
1	$\frac{1}{4}$	0	0	0
2	$\frac{1}{8}$	$\frac{1}{8}$	0	0
3	$\frac{1}{12}$	$\frac{1}{12}$	$\frac{1}{12}$	0
4	$\frac{1}{16}$	$\frac{1}{16}$	$\frac{1}{16}$	$\frac{1}{16}$
$q(y)$	$\frac{25}{48}$	$\frac{13}{48}$	$\frac{7}{48}$	$\frac{1}{16}$
$yq(y)$	$\frac{25}{48}$	$\frac{26}{48}$	$\frac{21}{48}$	$\frac{12}{48}$

$$\sum yq(y) = \tfrac{84}{48} = 1.75 = E(Y)$$

Under· certain circumstances the random variables X and Y are *independent*. Intuitively, this means that if the value of $X(s)$ is known, the distribution of values for $Y(s)$ are unaffected, and vice versa. For example, if 2 dice are tossed and if $X(s) =$ number on the first die while $Y(s) =$ number on the second die, then X and Y are intuitively independent. The formal definition of independence brings in the notion of independent events.

41 Definition

Let X and Y be random variables on a sample space S. X and Y are said to be *independent* if, for every possible value x of X and every possible value y of Y, the events $X = x$ and $Y = y$ are independent events.

Equivalently, let $A(x)$ be the set of s where $X(s) = x$ and let $B(y)$ be the set of s where $Y(s) = y$. Then X and Y are independent provided that

$$p(A(x) \cap B(y)) = p(A(x)) \cdot p(B(y)) \tag{5.30}$$

In fact, Equation 5.30 is precisely the condition that $A(x)$ and $B(y)$ be independent events (see Definition 31 and Equation 23 of Chapter 3).

We may rephrase Equation 5.30 using distributions.

42 Theorem

Let X and Y be random variables with joint distributions $p(x, y)$. Let X have distribution $p(x)$ and let Y have distribution $q(y)$. Then X and Y are *independent if and only if*

$$p(x, y) = p(x)q(y) \qquad \text{for all } x \text{ and } y \tag{5.31}$$

Proof. Using Definitions 9 and 37, we have

$$p(x, y) = p(A(x) \cap B(y)) \qquad p(x) = p(A(x)) \qquad q(y) = p(B(y))$$

Thus Equation 5.30 is equivalent to Equation 5.31.

43 Corollary

The joint distribution $p(x, y)$ of X and Y determines whether X and Y are independent or not.

In fact, we have seen in Theorem 39 that the distribution $p(x)$ of X and $q(y)$ of Y is determined from the joint distribution of X and Y. Thus it is possible to verify Equation 5.31, knowing only the values $p(x, y)$.

44 Example

X and Y are random variables whose joint distribution is given in Table 5.24. Determine whether X and Y are independent.

5.24 Joint Distribution

x \ y	3	4	5
0	.02	.04	.04
1	.08	.16	.16
2	.10	.20	.20

Method. We find $p(x)$ by summing rows and $q(y)$ by summing columns, then verify if $p(x, y) = p(x)q(y)$. In fact, this is seen to be the case in Fig. 5.25, and therefore X and Y are independent. The special case $p(1, 3) = p(1)q(3)$ is illustrated: $.08 = (.4)(.2)$.

5.25

x \ y	3	4	5	Sum
0	.02	.04	.04	.1
1	.08	.16	.16	.4
2	.10	.20	.20	.5
Sum	.2	.4	.4	

$$p(x, y) = p(x)q(y)$$

Thus the independence of X and Y can be inferred from Theorem 42 as in Example 44. However, there are better ways, in practice, to infer the independence of random variables. A most useful method occurs when a sample space is a product space $S \times T$ and the value $X(s, t)$ depends only on s, while $Y(s, t)$ depends only on t:

$$X(s, t) = X(s) \qquad Y(s, t) = Y(t)$$

In this case Theorem 51 of Chapter 3 shows that X and Y are independent. For example, suppose that a number X is determined by the result of an experiment. Let the experiment be repeated (independently) twice. Let X_1 be the number determined by the first experiment and let X_2 be determined by the second experiment. Then X_1 and X_2 are independent random variables. This technique will be useful in Chapter 6 when we analyze, theoretically, the notion of independent repetitions of an experiment.

The following theorem computes $E(XY)$ from the joint distribution of X and Y.

45 Theorem

Let X and Y be random variables on sample space S with joint distribution $p(x, y)$. $(x = x_1, \ldots, x_m; y = y_1, \ldots, y_n.)$ Then

$$E(XY) = \sum_{x,y} xyp(x, y) \tag{5.32}$$

$$E(XY) = \sum_y \sum_x xyp(x, y) \tag{5.33}$$

$$E(XY) = \sum_x \sum_y xyp(x, y) \tag{5.34}$$

Proof. We first note the meaning of the equations. In each equation the value $xyp(x, y)$ is computed for each possible value of x and y. Thus the values $xyp(x, y)$ are found for all points in the rectangular array as in Fig.

5.26 *Rectangular Array*

5.26. Equation 5.32 gives $E(XY)$ as the sum of all these values. Equation 5.33 computes this sum by finding the sum for each column $y(\Sigma_x xyp(x,y))$, and then summing these results $(\Sigma_y\Sigma_x xyp(x,y))$. Similarly, Equation 5.34 sums each row and then sums the results. Such sums as in Equation 5.32 through 5.34 are often called *double sums*, because they involve sums of sums.

To prove the result, we go to the definition of expectation. We have

$$E(XY) = \sum_{s \in S} p(s)X(s)Y(s)$$

We now break this sum into mn parts, one for each choice of x and y:

$$E(XY) = \sum_{x,y} \sum_{\substack{X(s)=x \\ Y(s)=y}} p(s)X(s)Y(s)$$

Each individual sum $\Sigma_{X=x,Y=y} p(s)X(s)Y(s)$ can be computed easily:

$$\sum_{\substack{X(s)=x \\ Y(s)=y}} p(s)X(s)Y(s) = \sum_{\substack{X=x \\ Y=y}} p(s)xy$$

$$= xy \sum_{\substack{X=x \\ Y=y}} p(s) = xyp(x,y)$$

Thus

$$E(XY) = \sum_{x,y} xyp(x,y)$$

This is the result.

We can now prove the following theorem, which will be useful for theoretical purposes in Chapter 6.

46 Theorem

Let X and Y be *independent* random variables on a sample space S. Then

$$E(XY) = E(X)E(Y) \tag{5.35}$$

Proof. We let $p(x)$ be the distribution of X and $q(y)$ the distribution of Y. Then, by Theorem 42, the joint distribution $p(x,y)$ of X and Y is given by

$$p(x,y) = p(x)q(y)$$

Using Theorems 45 and 19,

$$E(XY) = \sum_y \sum_x xyp(x,y)$$

$$= \sum_y \left(\sum_x xyp(x)q(y) \right)$$

$$= \sum_y \left(yq(y) \sum_x xp(x) \right)$$

$$= \sum_y yq(y)E(X)$$

$$= E(X) \sum_y yq(y)$$

$$= E(X)E(Y)$$

The reader should reconsider Example 26, where it was observed that $E(XY) \neq E(X)E(Y)$ in general.

EXERCISES

1. X and Y are defined according to the following table:

s	a	b	c	d	e	f
$p(s)$	.1	.3	.1	.2	.1	.2
$X(s)$	1	2	3	3	2	1
$Y(s)$	0	0	1	1	0	1

Construct a table for $p(x, y)$, the joint distribution of X and Y. Use this table to compute $p(x)$ and $q(y)$, the distribution of X and Y. Are X and Y independent? Explain.

2. Let X and Y have a joint distribution $p(x, y)$ in the following table:

x \ y	0	1	2	3
-1	.10	.20	.05	.10
0	.05	.05	0	.10
1	.20	0	.10	.05

  a. What is the distribution of X?
  b. What is the distribution of Y?
  c. Are X and Y independent?
  d. Find the probability that $X = Y$.
  e. Find the probability that $X < Y$.

f. Find $E(X)$, $E(Y)$, and $E(XY)$.

g. Find $E(X^2Y)$.

3. If $E(XY) = E(X)E(Y)$, are X and Y independent? Explain.

4. In the following (incomplete) table for the joint distribution $p(x, y)$ of X and Y, it is known that X and Y are independent. Complete this table.

x \ y	40	50	60
.1	.05	.25	.20
.2			.12
.3		.10	

5. A whole number is chosen at random from 1 through 10. If it is even, another number is chosen, but if it is odd, the play stops. Let X be the first number chosen and let Y be the number of cards chosen (1 or 2). Construct a table for the joint distribution of X and Y.

6. A number n is chosen at random between 1 and 10 inclusive. Let $D = D(n) = $ the number of positive divisors of n, and let $P = P(n) = $ the number of prime divisors of n. [Note that 1 is not considered a prime, so $P(1) = 0$.] Construct a table for the joint distribution $p(d, p)$ of D and P.

7. Prove that if X and Y have joint distribution $p(X, Y)$, then

$$E(X+Y) = \sum_{(x,y)} \sum (x+y)p(x, y)$$

$$E(X^2Y - Y^3) = \sum_{(x,y)} \sum (x^2y - y^3)p(x, y)$$

More generally, prove that if $f(x, y)$ is any function of x and y, then

$$E[f(X, Y)] = \sum \sum f(x, y)p(x, y)$$

8. Prove: If $X = $ constant and Y is an arbitrary random variable, then X and Y are independent.

9. Prove: If $Y = X^2$, then X and Y are independent if and only if $Y = $ constant.

10. Prove that if the joint distribution $p(x, y)$ of 2 independent random variables is arranged in the usual tabular form, then the rows are all proportional. Prove, conversely, that if the rows of the joint distribution $p(x, y)$ are proportional, then X and Y are independent.

*7 INTRODUCTION TO GAME THEORY

This section concerns an interesting and surprising application of probability to the theory of games. Although the games in this section will be comparatively simple, the analysis is very similar for much more complicated games.

Suppose that Arthur (player A) plays Bernard (player B) the following game: Each player chooses a head (H) or a tail (T) without knowledge of the other's choice. If the choices do not match, A loses \$1 to B. If they match (HH), A wins \$2. If they match (TT) A wins \$1. How should A play? How should B play?

This game is simply described by the *payoff matrix* in Table 5.27, which indicates A's winnings in all cases. (B loses the same amount, or wins the negative amount.) A pretheoretical analysis leads to the following thoughts: "A should play H because he wins more. But then B, figuring this out, should play T to take advantage of A's greed. But then A, sensing this, should play T, to doublecross B. But B, figuring this out, should play H. But then A should ...." This sort of infinite outsmarting leads nowhere. How then *should A play?*

5.27 A's Winnings (Payoff Matrix)

A \ B	H	T
H	2	−1
T	−1	1

Suppose A decides to play in the most conservative manner—that is, suppose A were to assume that B knows his (A's) move. Then, clearly regardless of A's move (H or T), A will lose \$1 (or win \$−1). Probability enters the picture if A decides to choose H and T with certain probabilities. In this way even A will not know what his move will be (until he plays). For example, suppose A chooses H and T, each with probability $\frac{1}{2}$ *and suppose A tells this to B.* (We say that A plays $\frac{1}{2}$H $+ \frac{1}{2}$T.) What does B do? B will presumably maximize his expected winnings (or, equivalently, he will minimize A's expected winnings). Whether B decides to play T or H, A's winnings are indicated in Fig. 5.28. A will win (on the average) \$0 or \$$\frac{1}{2}$ according as B plays T or H. Thus, from a position where A must lose \$1, he has arrived at a position where A will expect to win \$0 or \$1/2. B will play T, of course, in this situation, and A will break even.

A can now consider other probabilities for H and T. In general, the situation when A chooses H and T with probability p and $1-p$, respectively, is

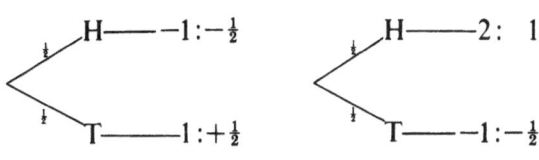

B plays T; $E(W) = 0$ B plays H; $E(W) = \frac{1}{2}$

5.28 *A's winnings if A plays $\frac{1}{2}H + \frac{1}{2}T$*

indicated in Fig. 5.29. [We say that A plays $pH + (1-p)T$.] Even if B knows A's strategy, all he can do is play T or H to minimize $1 - 2p$ and $-1 + 3p$. The smaller of $1 - 2p$ and $-1 + 3p$ is indicated by the dark, broken line in Fig. 5.30. It is found by graphing the 2 lines $y = 1 - 2p$ and $-1 + 3p$ and, for

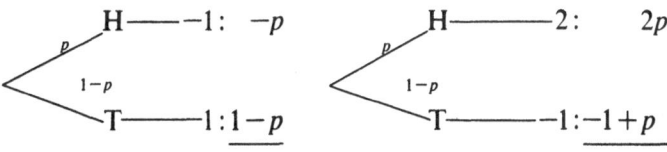

B plays T; $E(W) = 1 - 2p$ B plays H; $E(W) = -1 + 3p$

5.29 *A's Winnings [A plays $pH + (1-p)T$]*

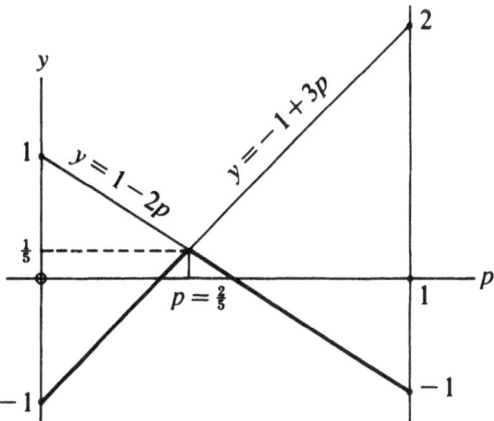

5.30 *A plays $pH + (1-p)T$; A's guaranteed winnings*

each p, choosing the least y value. If A plays conservatively, he will choose $p = \frac{2}{5}$, thereby guaranteeing himself an expected winning of $\frac{1}{5}$, regardless of how B plays. For any other choice of p, A can expect to win less if B plays properly. Thus A plays $\frac{2}{5}H + \frac{3}{5}T$ and will win (on the average) \$1/5 regardless of B's move. (Interestingly, we can also say that A will play T, most likely, but plays H occasionally "to keep B honest." However, the above analysis does not ascribe any such motivations to A. A merely wants to maximize his guaranteed expected winnings.)

Now the above analysis is from A's point of view. We have seen that A can guarantee himself an average win of \$.20 per game. What about B? He can reason similarly. Suppose B played most conservatively and announced to A that he intends to play $qH + (1-q)T$. In Fig. 5.31, we can compute A's winnings on the assumption that A plays H or T. If B announces his move, B will assume that A will maximize A's winnings and will play T or H according as $1-2q$ or $-1+3q$ is *larger*. We thus construct Fig. 5.32, which is similar to Fig. 5.30. Here, regardless of how B plays, A can win \$1/5 on the average. But if B plays $\frac{2}{5}H + \frac{3}{5}T$, A's expected winnings will be \$1/5, regardless of how A plays.

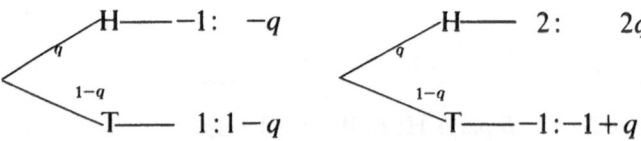

A plays T; $E(W) = 1-2q$ A plays H; $E(W) = -1+3q$

5.31 A's Winnings [B plays $qH + (1-q)T$]

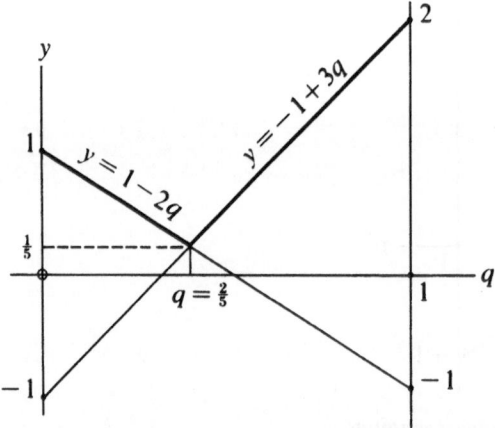

5.32 B plays $qH + (1-q)T$; A's Maximal Winnings

We summarize the situation. We allow both A and B to play H and T with any probabilities. If A plays $\frac{2}{5}H + \frac{3}{5}T$, nothing B does will effect A's winnings:

A will win \$1/5 on the average. If B plays $\frac{2}{5}H + \frac{3}{5}T$, nothing A does will effect A's winnings. A will win \$1/5 on the average. By introducing probability, we have given both A and B strategies, and we see that the value of this game to A is \$1/5.

The *theory of games* considers very broad generalizations of this procedure. We now state, without proof, a main result of that theory. Suppose A and B play a game in which A must choose any one of m strategies $s_1, \ldots, s_m$, while B must choose any of n strategies $t_1, \ldots, t_n$. If A chooses s_i and B chooses t_j, then A wins a_{ij} while B wins $-a_{ij}$. (The matrix $[a_{ij}]$ is called the *payoff matrix* of the game.) A and B are to choose their respective strategies without knowledge of the other's choice. Under these conditions, there is a value V (called the *value of the game*) with the following properties: A can choose strategy s_i with probability p_i ($\Sigma\, p_i = 1$) in such a way that, regardless of B's strategy, A's expected winnings will be V or *more*. B can choose strategy t_j with probability q_j ($\Sigma\, q_j = 1$) in such a way that, regardless of A's strategy, A's expected winnings will be V *or less*.

We illustrate this result with one more example.

47 Example

A and B play a game in which each chooses H or T without knowledge of the other's choice. A wins according to the following payoff matrix:

A \ B	H	T
H	2	1
T	0	5

B loses the amount that A wins. What is A's correct strategy? On the average, what can A expect to win?

Method. If A plays $pH + (1-p)T$, we can compute A's winnings as follows:

$$B \text{ plays H: } A \text{ wins } p \cdot 2 + (1-p) \cdot 0 = 2p$$
$$B \text{ plays T: } A \text{ wins } p \cdot 1 + (1-p)5 = 5 - 4p$$

Graphing $y = 2p$ and $y = 5 - 4p$, we obtain Fig. 5.33. The graphs intersect where $2p = 5 - 4p$. Thus $p = \frac{5}{6}$ and $y = 2p = \frac{10}{6} = 1\frac{2}{3}$. The dark, broken line indicates A's guaranteed winnings. If A plays H with probability $\frac{5}{6}$ and T with probability $\frac{1}{6}$ (the bluff!), A's expected winnings will be $1\frac{2}{3}$ — the value of the game.

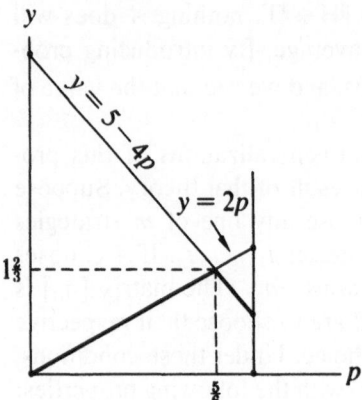

5.33 *A's Guaranteed Winnings When A Plays pH + $(1-p)$T.*

How should B play? If he plays qH + $(1-q)$T, then A's winnings are as follows:

A plays H: A wins $q \cdot 2 + (1-q) \cdot 1 = 1 + q$
A plays T: A wins $q \cdot 0 + (1-q) \cdot 5 = 5 - 5q$

A's maximal winnings are indicated in Fig. 5.34. The graphs intersect when $1 + q = 5 - 5q$ or $q = \frac{2}{3}$ and $y = 1 + q = 1\frac{2}{3}$, as before. The dark, broken line indicates A's maximal winnings. If B plays H with probability $\frac{2}{3}$ and T with probability $\frac{1}{3}$ (B's bluff!), A's expected winnings will be $1\frac{2}{3}$—the value of the game.

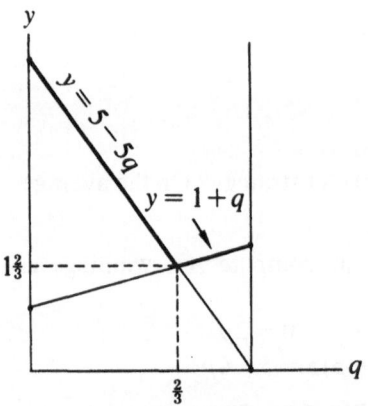

5.34 *A's Maximal Winnings When B Plays qH + $(1-q)$T.*

It may be argued that the games considered are so childish, dull, and uninteresting that they do not deserve to be designated games. However, we point out that these are only the simplest games and are used because they are capable of being analyzed simply. An argument can be made that many two-person, zero-sum games can be put into the form considered on page 197. (A zero-sum game is one in which one person loses what another wins — there is no outside source of money that will make them "cooperate.") To see this, imagine any game (blackjack, poker, Monopoly, chess, . . .). Then imagine writing down all strategies. These strategies tell anyone what you will do in any circumstance. Admittedly, each strategy will tend to be book length, and there will be an enormous number of strategies. However, once these strategies are listed $(s_1, \ldots, s_m)$, all you do to play the game is to choose one strategy s_i! Thus we are back in the simple situation described on page 197.

For an introduction to game theory, written on an elementary level, the reader is referred to *The Compleat Strategyst* by John D. Williams (New York, McGraw-Hill, 1954).

EXERCISES

1. Compute the value of the game and A's correct strategy if A's payoff matrix is

A \ B	H	T
H	1	−1
T	−1	10

Also compute B's strategy.

2. As in Exercise 1, compute the value and strategies for the payoff matrix

$$\begin{bmatrix} -1 & 3 \\ 2 & -1 \end{bmatrix}$$

3. As in Exercise 1, compute the value and strategies for the payoff matrix

$$\begin{bmatrix} 1 & 4 \\ 3 & 5 \end{bmatrix}$$

(*Hint:* Think about this game before writing any equations!)

4. As in Exercise 1, compute the value and strategies for the payoff matrix

$$\begin{bmatrix} 1 & -2 \\ -2 & 4 \end{bmatrix}$$

5. A and B play the following game. Each puts out 1 or 2 fingers. B pays A the sum of the fingers. However, if both put out 2 fingers, then A pays B the sum (4) of the fingers. What should A do? What should B do? How much can A expect to win?

6. A and B play a game with the following payoff matrix for A:

A \\ B	t_1	t_2	t_3
s_1	10	0	5
s_2	0	10	2

a. What should A do? How much can A expect to win?
b. What should B do? (*Hint:* Look at A's best strategy. If you were B, you would not play any strategy that gives A more than he deserves.)

*8 SYMMETRY IN RANDOM VARIABLES

In Section 4 of Chapter 4 we discussed the use of symmetry in computing probabilities. We shall now show how symmetry considerations can be used in the computation of expected values. We recall that a transformation $s \to s'$ of S into S is called a *symmetry* if each point of S is equal to s' for one and only one value of s, and if $p(s) = p(s')$ (cf. Definitions 23 and 24 of Chapter 4). If X is a random variable on S, we define a random variable X' by the formula

$$X'(s) = X(s') \tag{5.36}$$

For example, suppose 10 coins are tossed and S is the appropriate sample space of 2^{10} elements. Let $s \to s'$ change heads to tails and tails to heads. (For example, HTTHH$\ldots \to$ THHTT$\ldots$) Then if $X =$ number of heads, $X' =$ number of tails. For $X'(s) = X(s') =$ number of heads in $s' =$ number of tails in s. Similarly, if $Y =$ excess of heads over tails, $Y' =$ excess of tails over heads. In Section 4 of Chapter 4 the fundamental result was that if A and A' were corresponding sets under a symmetry, then A and A' had the same probability. We now state and prove the analogous result for random variables.

48 Theorem

Let $s \to s'$ be a symmetry of S, let X be a random variable on S, and let X' be the random variable defined by Equation 5.36. Then

$$E(X) = E(X') \qquad (5.37)$$

Proof. We use the fact that as s runs through S, so does s'.

$$E(X') = \sum_s p(s)X'(s) \qquad \text{(definition)}$$

$$= \sum_s p(s)X(s') \qquad \text{(Equation 5.36)}$$

$$= \sum_s p(s')X(s') \qquad (s \to s' \text{ is a symmetry})$$

$$= \sum_s p(s)X(s) \qquad \text{(remark above)}$$

$$= E(X) \qquad \text{(definition)}$$

We now show how Theorem 48 can be exploited in the computation of some probabilities.

49 Example

Ten fair coins are tossed. On the average, how many heads turn up?

Method. Intuitively, since a head and a tail are equally likely, we may expect an equal number of heads as tails. Hence, on the average, there are 5 heads (and 5 tails). We now formalize this reasoning.

Let X = number of heads. (We are taking the usual sample space of 2^{10} elements.) Then, if $s \to s'$ interchanges H and T, X' = number of tails. But

$$\text{number of heads} + \text{number of tails} = 10$$

or

$$X + X' = 10 \qquad (5.38)$$

Taking expected values,

$$E(X) + E(X') = E(10) = 10$$

Since $E(X') = E(X)$ by Theorem 48, we have

$$2E(X) = 10$$

$$E(X) = 5$$

Thus there are 5 heads on the average.

50 Example

A whole number N is chosen at random from 1 through 100. What is the average value of N?

Method. We can do this directly from definition. If $k = 1, 2, \ldots, 100$, $N = k$ with probability $\frac{1}{100}$. Thus

$$E(N) = \tfrac{1}{100} \cdot 1 + \tfrac{1}{100} \cdot 2 + \cdots + \tfrac{1}{100} \cdot 100$$

$$= \tfrac{1}{100}(1 + 2 + \cdots + 100)$$

$$= \frac{1}{100} \cdot \frac{100 \cdot 101}{2} = 50.5$$

Here we have used the formula for summing an arithmetic progression:

$$1 + 2 + \cdots + n = \frac{n(n+1)}{2}$$

We now prove that $E(N) = 50.5$ by symmetry considerations. Corresponding to any number n, we determine $n' = 101 - n$. Thus $n \to n'$, and we have $1 \to 100$, $2 \to 99, \ldots$. This is a symmetry. Since $N(n) = n$, we have

$$N'(n) = N(n') = N(101 - n) = 101 - n = 101 - N(n)$$

Thus

$$N' = 101 - N$$

Taking expected values and using $E(N') = E(N)$, we obtain

$$E(N) = 101 - E(N)$$

Thus $2E(N) = 101$ and $E(N) = 50.5$.

51 Example

Three dice are tossed. Let L = lowest number turning up, H = highest number turning up, M = middle number turning up. Prove that $E(H) + E(L) = 7$ and that $E(M) = 3.5$.

Method. Before proving this formally let us note that it is intuitively obvious.

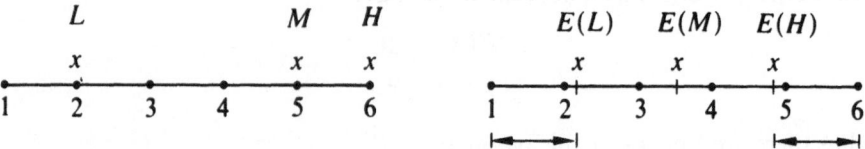

5.35   *Outcome of 3 Dice, with Expectations*

In the left half of Fig. 5.35 we have a possible outcome of a toss. If averaging occurs, we may (intuitively) expect $E(L), E(H), E(M)$ as in the right half of Fig. 5.35. Thus, since M is just as likely to fall low as high, we split the difference and expect $E(M) = (\frac{1}{2})(1+6) = 3.5$. Also, if $E(L)$ is h above 1, we (intuitively) expect $E(H)$ to be h below 6 "by symmetry." Thus $E(L) = 1+h$, $E(H) = 6-h$, and therefore $E(L)+E(H) = 7$. Thus the only problem is to translate intuitive symmetry into concrete (mathematical) terms.

The trick is to change high into low by the correspondence

$$\begin{array}{cccccc} 1 & 2 & 3 & 4 & 5 & 6 \\ \downarrow & \downarrow & \downarrow & \downarrow & \downarrow & \downarrow \\ 6 & 5 & 4 & 3 & 2 & 1 \end{array}$$

Thus, if n is a point of a die, $n' = 7-n$. (In Fig. 5.35 this reflects points about the value 3.5.) Now suppose 3 dice are tossed, with sample point $s = (n_1, n_2, n_3)$. Then

$$s' = (7-n_1, 7-n_2, 7-n_3)$$

We have high point going into low point, middle into middle, and low into high. Thus

$$L'(s) = L(s') = \text{lowest of } (7-n_1, 7-n_2, 7-n_3)$$

$$= 7 - \text{highest of } (n_1, n_2, n_3)$$

$$L'(s) = 7 - H(s) \tag{5.39}$$

Similarly,

$$M'(s) = M(s') = \text{middle of } (7-n_1, 7-n_2, 7-n_3)$$

$$= 7 - \text{middle of } (n_1, n_2, n_3)$$

$$M'(s) = 7 - M(s) \tag{5.40}$$

Now take expected values in Equations 5.39 and 5.40, using Theorem 48. We obtain

$$E(L) = 7 - E(H)$$

$$E(M) = 7 - E(M)$$

which are the results.

We conclude this chapter with an illustration that involves a less obvious symmetry and yields a more interesting result.

52 Example

A deck of cards is well shuffled. On the average, how many cards are on top of the first ace from the top? (Cf. Exercise 13 of Section 4.4.)

Method. The sample space of 52! elements is enormous. However, for any arrangement s we have 4 aces, which divide the rest of the deck into 5 parts, consisting of X_1 cards on top of the first ace, X_2 cards between the first and second ace, etc. (See Fig. 5.36.) Now we can interchange the roles

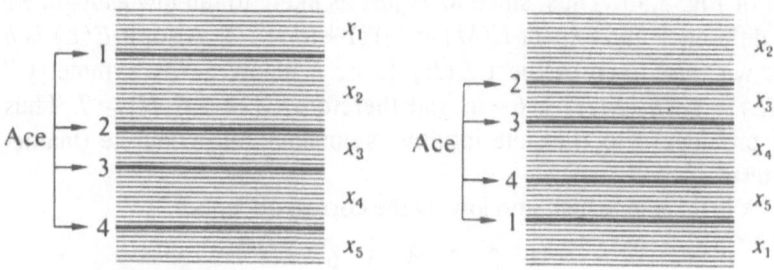

5.36   Changing the Deck

of $X_1, X_2, \ldots, X_5$ by changing the deck s into a new deck s' as follows: Take all the cards on top of the first ace and put them on the bottom of the deck (same order) with the first ace now on top of these X_1 cards. A glance at Fig. 5.36 shows that $s \rightarrow s'$ is a symmetry. In fact, every arrangement t of a deck comes from one and only one arrangement s ($t = s'$ has only one solution s). Also $p(s) = p(s') = 1/52!$ Clearly (see Fig. 5.36),

$$X_1'(s) = X_1(s') = X_2(s)$$

$$X_2'(s) = X_2(s') = X_3(s)$$

$$\vdots$$

$$X_5'(s) = X_5(s') = X_1(s)$$

Thus $E(X_1') = E(X_2)$, $E(X_2') = E(X_3)$, etc. But by Theorem 48 this implies that

$$E(X_1) = E(X_2) = E(X_3) = E(X_4) = E(X_5) \qquad (5.41)$$

But there are a total of 48 cards exclusive of aces. Thus

$$X_1 + X_2 + \cdots + X_5 = 48$$

Taking expected values and using Equation 5.41, we obtain

$$5E(X_1) = 48$$

and, finally,

$$E(X_1) = \tfrac{48}{5} = 9.6$$

Intuitively, the 48 cards split up — on the average — into 5 equal groups of 9.6 each.

The reader should note the common theme of Example 49 through 52. In all cases there were several random variables and a relationship among them. (In Example 52, $X_1, X_2, \ldots, X_5$ were the random variables, and $X_1 + \cdots + X_5 = 48$ was the relationship.) Then there was a symmetry (whose existence, let it be said, comes through a combination of intuition, experience, and work). Finally, the symmetry induced some connection between the random variables, which, together with Theorem 48 and the given relationship, solved the problem – or as much of the problem as the symmetry could.

EXERCISES

1. If n fair coins are tossed, prove by symmetry considerations that the expected number of heads is $n/2$.

2. If n dice are tossed, prove by symmetry considerations that the expected number of 3's is $n/6$. (*Hint:* Symmetry tells us that the expected number of 3's = expected number of 4's = etc.)

3. Suppose 11 coins are tossed. By symmetry considerations, show that the expected excess of heads over tails is 0.

4. One hundred people are each asked to think of a number 1 or 2. Assume that they do this independently and that 1 and 2 are equally likely. Using symmetry considerations, prove that the expected number of people who choose 1 is 50.

5. An urn contains 9 white balls and 1 black ball. Balls are successively chosen at random and without replacement until the black ball is chosen. On the average, how many white balls will be chosen before the black ball is reached?

6. Do Exercise 5 on the assumption that the urn contains 9 white and 2 black balls. Generalize.

7. Five dice are tossed. Let $Y =$ largest number tossed plus smallest number tossed. Prove that $E(Y) = 7$.

8. A deck of cards is well shuffled. Cards are dealt off the top until a black card appears. On the average, how many cards are dealt? (Include the black card.)

CHAPTER 6 STANDARD DEVIATION

INTRODUCTION

In Chapter 5 we defined the mean $\mu = E(X)$ as an important single number associated with a random variable X. It was pointed out that the mean cannot replace X, because it gives no indication of how the values $X(s)$ are distributed about the mean. In this chapter we introduce the numbers $V(X)$, the *variance* of X, and $\sigma(X) = \sqrt{V(X)}$, the *standard deviation*, either of which is a measure of the "spread" of X about its mean. Figure 6.1 shows the graphs of three distributions, each with the same mean μ. It is seen (intuitively) that the "spread" of $p_1(x)$ is greater than that of $p_2(x)$, which in turn has a greater spread than $p_3(x)$. In the next section we shall learn how to measure this spread (standard deviation) exactly.

Experimentally, none of the observed values of X need be near the mean

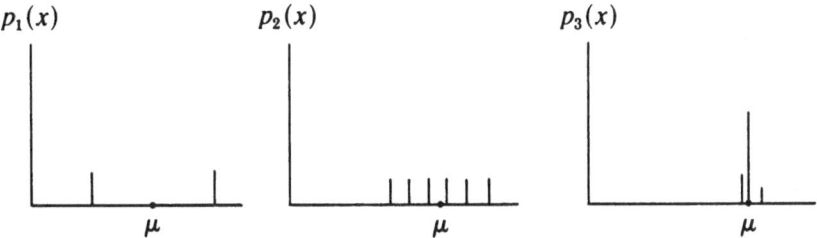

6.1 Some Distributions

$\mu = E(X)$. The standard deviation will be a measure of how far the values of X are from μ. Since $X - \mu$ is the deviation of X from u, it might appear that the average value of $X - \mu$, $E(X - \mu)$, is a measure of the deviation. However,

$$E(X - \mu) = E(X) - E(\mu) = \mu - \mu = 0$$

Thus $X - \mu$ is sometimes positive, and sometimes negative, and these values have 0 average. The simplest *nonnegative* measure of the deviation of X from μ is $(X - \mu)^2$, and this will be the basis of our definition in Section 1.

In Section 5 we shall show how, with the help of the standard deviation, we can relate probability to long-term relative frequency as in our original motivation of probability.

1 VARIANCE AND STANDARD DEVIATION

If X is a random variable, we shall now consistently use the symbol μ to designate its mean. Occasionally, if the dependence on X is to be made explicit, we use μ_x. Thus

$$E(X) = \mu_x = \mu \tag{6.1}$$

1 Definition
Let X be a random variable. The *variance* of X, written $V(X)$, is defined by the formula

$$V(X) = E((X - \mu)^2) \tag{6.2}$$

This may be written explicitly in terms of X:

$$V(X) = E((X - E(X))^2) \tag{6.3}$$

As noted in the Introduction, $V(X)$ is a measure of how much X varies from its mean μ. We illustrate with a numerical example.

2 Example
Let X be defined as follows:

s	s_1	s_2	s_3
$p(s)$	.2	.3	.5
$X(s)$	1	2	4

Find $\mu = E(X)$ and find $V(X)$.

Method. We first find $\mu = E(X) = \Sigma p(s)X(s) = (.2)(1) + (.3)(2) + (.5)(4) = 2.8$. We then compute the value of $X(s) - \mu = X(s) - 2.8$ and square to find $[X(s) - 2.8]^2$. We then find the variance of X, which is the expected value of this random variable: $V(X) = E((X - 2.8)^2) = \Sigma p(s)[X(s) - 2.8]^2$. The result is $V(X) = 1.56$. The computation is conveniently done in Table 6.2. We shall later learn how to simplify the calculation somewhat, although some computation will still be necessary. For a large table, a desk computer or slide rule will prove useful, even necessary. For many calculations of this kind, high-speed digital computers are used.

6.2 Computation of V(X)

s	p	X	pX	$X-\mu$	$(X-\mu)^2$	$p(X-\mu)^2$
s_1	.2	1	.2	−1.8	3.24	.648
s_2	.3	2	.6	−.8	.64	.192
s_3	.5	4	2.0	1.2	1.44	.720
Sum			$\overline{2.8}\ (=\mu)$			$\overline{1.560}\ [=V(X)]$

We now prove some basic properties of the variance.

3 Theorem

Let X be a random variable with mean μ and let c be a constant. Then

$$V(cX) = c^2 V(X) \tag{6.4}$$

$$V(X+c) = V(X) \tag{6.5}$$

$$V(c) = 0 \tag{6.6}$$

$$V(X) > 0 \qquad \text{if } X \neq \text{constant} \tag{6.7}$$

Proof. Since $E(X) = \mu$, we have $E(cX) = cE(X) = c\mu$. Thus $V(cX)$ may be found from Equation 6.3:

$$V(cX) = E((cX - c\mu)^2)$$
$$= E(c^2(X - \mu)^2)$$
$$= c^2 E((X - \mu)^2)$$
$$= c^2 V(X)$$

This proves Equation 6.4.

We have $E(X+c) = E(X) + c = \mu + c$. Thus

$$V(X+c) = E((X+c) - (\mu+c))^2$$
$$= E((X-\mu)^2)$$
$$= V(X)$$

This proves Equation 6.5.

The proof of Equation 6.6 is a direct consequence of the definition. Since $E(c) = c$, we have

$$V(c) = E((c-c)^2) = E(0) = 0$$

Finally,

$$V(X) = \Sigma p(s)[X(s) - \mu]^2$$

is the sum of nonnegative terms. If $X(s)$ is not a constant, one of the terms $X(s) - \mu$ is not equal to 0. Thus $p(s)[X(s) - \mu]^2 > 0$ for the summand, and the entire sum is therefore positive. This completes the proof of the theorem. Strictly, 6.7 is true only if all sample points have positive probability. This is seen by the proof, and it is in this sense that we use the inequality 6.7.

On page 177 we noted that the significance of the equation $E(cX) = cE(X)$ was that the mean $\mu = E(X)$ had the same *units* as X. Equation 6.4 shows that $V(X)$ is measured in *square units*. Thus, if X is measured in feet, pounds, or years, $V(X)$ will be measured in square feet, square pounds, or square years. For example, suppose $X(s)$ is the height of student s in feet. Then we claim that $V(s)$ is measured in square feet. We can see this by changing units. For example, $Y = 12X$ is the height in inches. Thus $V(Y) = V(12X) = 144V(X)$, and, therefore, if we convert X to inches by multiplying by 12, $V(X)$ will be converted to square inches (by multiplying by 144). A similar remark applies to any change of units.

Equation 6.5 shows that $V(X)$ is unchanged if X is changed to $X+c$. This, in turn, shows that the variance is not affected by a uniform shift in the value of X. In Fig. 6.3 the distributions pictured are distributions for X and $X+c$. It is reasonable that the variance, as a measure of spread, should be the same for both distributions.

$p_1(x)$

$p_2(x)$

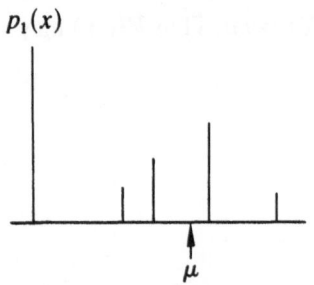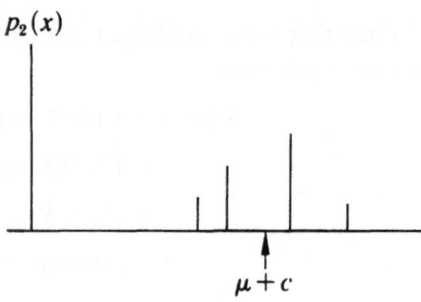

μ $\qquad\qquad\qquad\qquad \mu+c$

6.3   Shifted Distribution

In order to have the same units, we choose the square root of $V(X)$ as the measure of the spread of X.

4 Definition

The *standard deviation* of X, written $\sigma(X)$, is defined by the formula

$$\sigma(X) = \sqrt{V(X)} \qquad (6.8)$$

We shall simply use σ to designate the standard deviation of X. If the dependence on X is to be made explicit, we use σ_x or $\sigma(X)$. Thus

$$\sigma(X) = \sigma_x = \sigma \qquad \sigma^2 = V(X) \qquad (6.9)$$

Taking positive square roots in Equations 6.4 through 6.7, we obtain the following results:

$$\sigma(cX) = |c|\sigma(X)^1 \qquad (6.10)$$

$$\sigma(X+c) = \sigma(X) \qquad (6.11)$$

$$\sigma(c) = 0 \qquad (6.12)$$

$$\sigma(X) > 0 \qquad \text{if } X \neq \text{constant} \qquad (6.13)$$

Equation 6.10 shows that σ has the same units as X. This allows us to picture σ geometrically, when the values of X are so pictured. In Example 2 we found that $V(X) = 1.56$. Thus $\sigma = \sqrt{1.56} = 1.25$.[2] The distribution of Example 2 is plotted in Fig. 6.4. In it, the mean and standard deviation are

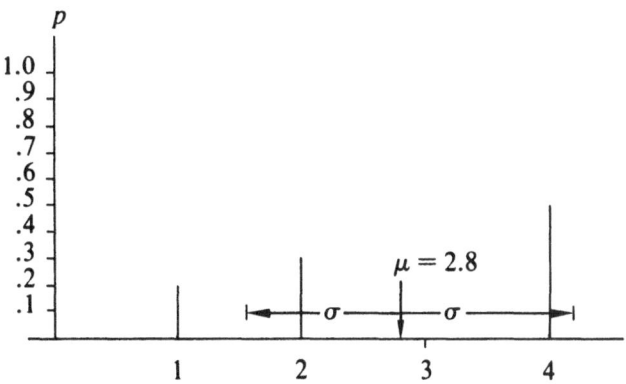

6.4 Distribution with Mean and Standard Deviation

1 $\sqrt{c^2} = |c|$. The equation $\sqrt{c^2} = c$ is not true for negative c. Thus $\sqrt{(-3)^2} = \sqrt{9} = 3$, not -3. By definition, $|c| = \pm c$, where the sign is chosen to make $|c|$ positive.
2 Square roots can be obtained easily with the aid of a slide rule. An alternative method is to use Appendix C. A slightly worse way is to use a table of logarithms. There is also an algorithm, which nobody remembers *or* uses.

indicated. Since the mean and standard deviation have the same units as X, the figure is not misleading. [If, for example, $V(X)$ were plotted as a length, the figure *would* be misleading, because the units would be wrong.]

A useful formula for $V(X)$ is obtained in the following theorem.

5 Theorem
Let X have mean μ. Then

$$\sigma^2 = V(X) = E(X^2) - \mu^2 \qquad (6.14)$$

or

$$\sigma^2 = V(X) = E(X^2) - [E(X)]^2 \qquad (6.15)$$

The proof uses the definition and properties of the expectation:

$$\begin{aligned}
V(X) &= E((X-\mu)^2) \\
&= E(X^2 - 2\mu X + \mu^2) \\
&= E(X^2) - 2\mu E(X) + \mu^2 \\
&= E(X^2) - 2\mu \cdot \mu + \mu^2 \\
&= E(X^2) - \mu^2
\end{aligned}$$

Equation 6.14 can sometimes be used to calculate the expected value of X^2. Thus

$$E(X^2) = \sigma^2 + \mu^2 \qquad (= \sigma_x^2 + \mu_x^2) \qquad (6.16)$$

Equation 6.14 is often used for computational purposes. Table 6.5 shows the work for the distribution of Example 2. (Cf. Table 6.2.)

6.5 Computation of V(X)

p	X	$pX \cdot$	X^2	pX^2
.2	1	.2	1	.2
.3	2	.6	4	1.2
.5	4	2.0	16	8.0
		2.8 $(=\mu)$		9.4 $[=E(X^2)]$

$$\sigma^2 = V(X) = E(X^2) - \mu^2 = 9.4 - (2.8)^2 = 1.56$$

If X has the distribution $p(x)$, we can find the variance directly from the distribution, much as we did in Tables 6.2 and 6.5. We state the formulas, reviewing the formula for the mean μ. The proof is as in the proof of Theorem 19 of Chapter 5, and is left to the reader.

If X has the distribution $p(x)$, $x = x_1, \ldots, x_k$, then

$$\mu = E(X) = \sum_{i=1}^{k} x_i p(x_1) \qquad (6.17)$$

$$\sigma^2 = V(X) = \sum_{i=1}^{k} (x_i - \mu)^2 p(x_i) \qquad (cf.\ 6.2) \qquad (6.18)$$

$$\sigma^2 = V(X) = \sum_{i=1}^{k} x_i^2 p(x_i) - \mu^2 \qquad (cf.\ 6.15) \qquad (6.19)$$

If an experiment is run N times, and values $x_i, \ldots, x_k$ are observed with frequency $n_1, \ldots, n_k$, then the above formulas may be used provided $p(x)$ is taken to be the relative frequency of x. Thus we replace $p(x_i)$ by the relative frequency $f_i = n_i/N$ of the result x_i. The numbers obtained are called the *sample mean* and the *sample variance* so that they will not be confused with the mean and the variance. (These latter numbers use probabilities in their computation: the former use observed relative frequencies.) The sample mean will be designated by m (sometimes $\bar{x}$ is used) and the sample variance will be denoted by s^2. The *sample standard deviation s* is defined as the positive square root of the sample variance. The formulas for computing m and s^2 follow from Equations 6.17 through 6.19. We list them here for convenience.

Suppose the value x_i occurs n_i times. Let $\Sigma n_i = N = $ total number of observations and $f_i = n_i/N$. Then[3]

$$m = \bar{x} = \sum x_i f_i = \frac{1}{N} \sum x_i n_i \qquad (6.20)$$

$$s^2 = \sum (x_i - m)^2 f_i = \frac{1}{N} \sum (x_i - m)^2 n_i \qquad (6.21)$$

$$s^2 = \sum x_i^2 f_i - m^2 = \frac{1}{N} \sum x_i^2 n_i - m^2 \qquad (6.22)$$

If the results of N experiments are $x_1, x_2, \ldots, x_N$, it is sometimes inconvenient to sort out the frequencies of each observation x_i. In that case we can use the following formulas. *Suppose that N experiments yield the observations $x_1, x_2, \ldots, x_n$, with possible repetitions. Then the sample mean and the sample variance are given by*

$$m = \bar{x} = \frac{1}{N} \sum_{i=1}^{N} x_i \qquad (6.23)$$

3 Some authors use $1/(N-1)$ instead of $1/N$ in 6.21 and 6.22. The reason for this will be explained in Section 4.

$$s^2 = \frac{1}{N} \sum_{i=1}^{N} (x_i - m)^2 \tag{6.24}$$

$$s^2 = \frac{1}{N} \sum_{i=1}^{N} x_i^2 - m^2 \tag{6.25}$$

(If like terms are combined, these equations lead to Equations 6.20 through 6.22.) Equations 6.23 through 6.25 follow directly from the definition of the mean and variance, provided we regard $x_1, \ldots, x_n$ as the N value of a random variable over a *uniform* sample space S. For, *in this case*,

$$E(X) = \sum p(s)X(s) = \sum_{i=1}^{N} \frac{1}{N} x_i = \frac{1}{N} \sum_{i=1}^{N} x_i = \bar{x}$$

and

$$V(X) = \sum p(s)[X(s) - E(X)]^2 = \frac{1}{N} \sum_{i=1}^{N} (x_i - \bar{x})^2 = s^2$$

We shall illustrate with two examples.

6 Example

The daily maximum temperature in New York City during June 1965 was as follows:

Date	Max. temp. (°F)	Date	Max. temp. (°F)	Date	Max. temp. (°F)
June 1	77	June 11	81	June 21	90
2	70	12	84	22	91
3	70	13	68	23	93
4	74	14	66	24	82
5	81	15	62	25	79
6	82	16	62	26	79
7	88	17	73	27	79
8	90	18	73	28	88
9	83	19	80	29	95
10	88	20	89	30	77

What is the sample mean and sample standard deviation?

Method. The numbers are sufficiently nonrepetitive, so we shall use Equations 6.23 through 6.25. If x is the temperature, it is more convenient to work with $y = x - 80$. This corresponds to using $Y = X - 80$ as a new random variable whose sample mean is $m - 80$ and whose sample variance is the same as that of X (see Equation 6.11). We compute as in the following table:

x	$y = x - 80$	y^2	x	$y = x - 80$	y^2
77	−3	9	73	−7	49
70	−10	100	73	−7	49
70	−10	100	80	0	0
74	−6	36	89	9	81
81	1	1	90	10	100
82	2	4	91	11	121
88	8	64	93	13	169
90	10	100	82	2	4
83	3	9	79	−1	1
88	8	64	79	−1	1
81	1	1	79	−1	1
84	4	16	88	8	64
68	−12	144	95	15	225
66	−14	196	77	−3	9
62	−18	324	Sum	−6	2,366
62	−18	324			

$$m_y = \frac{1}{N} \sum y_i = \left(\tfrac{1}{30}\right)(-6) = -.20$$

$$y = x - 80 \qquad x = y + 80$$

Thus $m_x = m_y + 80 = 79.80$.

$$s_y^2 = \frac{1}{N} \sum y_i^2 - m_y^2$$

$$= \tfrac{1}{30} \cdot 2{,}366 - .04$$

$$= 78.87 - .04 = 78.83$$

Therefore, $s_x^2 = s_y^2 = 78.83$.

$$s_x = \sqrt{78.83} = 8.88$$

Thus, to 1 decimal place,

$$m = 79.8°$$

$$s = 8.9°$$

The calculations show that the average maximum daily temperature in New York City during June 1965 was 79.8°. The sample standard deviation was $s = 8.9°$. The residents of the city were as much interested in the sample mean as in the standard deviation. Note that $79.8° + 8.9° = 88.7°$, a temperature exceeded on 6 days out of 30 (relative frequency $\frac{6}{30} = 20$ percent). Similarly, $79.8° - 8.9° = 70.9°$. The maximum temperature was less than this also for 6 days. *In this case*, the maximum temperature fell in the range $m - s$ to $m + s$ 60 percent of the time. Also, *all* temperatures fell in the range $m - 2s$ to $m + 2s$. We shall have more to say about how accurately the standard deviation (defined in its mysterious way) serves to determine the range in which the values $X(s)$ fall.

7 Example

In Table 1.1, in which the highest number of 3 dice was recorded, compute the sample mean and sample standard deviation.

Method. The following table is self-explanatory. We use Equations 6.20 through 6.22, because the frequencies are tabulated. The sample mean is 4.99 (cf. Example 16 of Chapter 5) and the sample standard deviation is 1.20.

x	n	nx	x^2	nx^2
1	1	1	1	1
2	6	12	4	24
3	5	15	9	45
4	12	48	16	192
5	33	165	25	825
6	43	258	36	1,548
Sum	$100 = N$	499		2,635

$$m = \frac{1}{N} \sum x_i n_i = \left(\tfrac{1}{100}\right)(499) = 4.99$$

$$s^2 = \frac{1}{N} \sum x_i^2 n_i - m^2 = \left(\tfrac{1}{100}\right)(2,635) - (4.99)^2 = 1.45$$

$$s = \sqrt{1.45} = 1.20$$

EXERCISES

1. A fair coin is tossed. Let $X = 1$ if H occurs and let $X = 0$ if T occurs. Compute the mean μ of X and the standard deviation σ.

2a. Two coins are tossed. Let $X =$ number of heads. Compute the mean μ of X and the standard deviation σ.

b. Find μ and σ if $X =$ number of heads when 3 coins are tossed.

3. Let $S = \{s, f\}$ be a 2-point sample space. Suppose that $p(s) = p$, $p(f) = q$, where $p + q = 1$. Let $X(s) = 1$, $X(f) = 0$. Compute μ and σ for X.

4. A random variable X has the following distribution table:

x	0.2	0.4	0.5
$p(x)$	.2	.7	.1

Find μ and σ for X.

5. The grades in a class were distributed as follows:

Grade	100	90	80	70	60	50
Number of students	2	5	8	7	2	3

Find the sample average grade and the sample standard deviation.

6. Let X be a random variable. Which is larger, $E(X^2)$ or $E(X)^2$? Why?

7. Find an expression for $E((X - \mu)^3)$ which involves $E(X^3)$, μ, and σ. (*Hint:* Use Equation 6.14.)

8. A factory produces precision spherical steel balls, 2 cm in diameter. A quality-control expert samples 100 of these balls and obtains the following data:

Diameter (cm)	1.97	1.98	1.99	2.00	2.01	2.02	2.03
Frequency	2	7	20	45	19	6	1

Compute the sample mean diameter and the sample standard deviation.

9. Let c be a constant, and let μ and σ be the mean and standard deviation of X. Prove:

$$E((X-c)^2) = \sigma^2 + (\mu-c)^2$$

(*Hint:* Use Equation 6.16, applied to $Y = X - c$.)

10. The number of triple plays per season during the 1930s is given in the following table:

1930: 4	1934: 3	1937: 3
1931: 2	1935: 2	1938: 0
1932: 2	1936: 2	1939: 2
1933: 2		

Find the sample mean number of triple plays and the sample standard deviation. Do the same for the number of triple plays per season during the 1940s, and compare your answers.

1940: 1	1944: 5	1947: 4
1941: 0	1945: 0	1948: 3
1942: 3	1946: 0	1949: 6
1943: 0		

11. Table 5.11 gives the probability that the high number of three dice is $1, 2, \ldots, 6$. Compute the mean high number and the standard deviation. Compare with the sample mean and the sample standard in Example 7.

2 CHEBYSHEV'S THEOREM

If a random variable X has mean μ and standard deviation σ, what information can be deduced concerning the distribution of the values of X? Intuitively, we should feel that the values of X are on either side of μ, and that a small σ ensures that the values of X are near μ with a large probability. For experimental results, a small mean standard deviation indicates that a large percentage of the samples are near the mean sample m. Chebyshev's theorem gives precise estimates concerning *what* percentage and *how* near.

The distance of $X(s)$ from the mean μ is $|X(s) - \mu|$. Absolute values are necessary to assure a positive answer regardless of whether $X(s)$ is smaller or larger than μ. Thus the equation $|X(s) - \mu| < k$ can be paraphrased "$X(s)$ is within k units of μ." Similarly, $|x - \mu| \geq k$ means "x is at least k units away from μ."

It is convenient to measure the distance of X from μ in units of σ. Thus we might ask for the probability that X is within 2σ of μ, or for the probability that X is more than 2.5σ away from μ. Chebyshev's theorem gives an estimate for the probability that X is within $z\sigma$ of μ. (z is any positive real number.)

8 Theorem. Chebyshev

If a random variable X has mean μ and standard deviation $\sigma > 0$, then the probability that $|X - \mu| < z\sigma$ is at least $1 - 1/z^2$.

Before proving this theorem, we give a few examples. For $z = 2$ we find that the probability that $|X - \mu| < 2\sigma$ is at least $1 - \frac{1}{4} = \frac{3}{4}$. Thus Chebyshev's theorem guarantees that X is strictly within 2σ of μ with probability $\geq \frac{3}{4}$. In practice, $\frac{3}{4}$ is often well exceeded, but the beauty of this theorem is its generality; it works for *all* random variables. For $z = 1$ we find that the probability that $|X - \mu| < \sigma$ is at least 0 (i.e., ≥ 0), a noninformative statement. Thus Chebyshev's theorem gives no information for $z = 1$ (and similarly for $z < 1$).

Proof. We let
$$p = \text{probability that } |X(s) - \mu| < z\sigma$$
Thus
$$1 - p = \text{probability that } |X(s) - \mu| \geq z\sigma$$
and, by definition
$$p = \sum_{|X(s) - \mu| < z\sigma} p(s) \qquad 1 - p = \sum_{|X(s) - \mu| \geq z\sigma} p(s) \qquad (6.26)$$
We now use the definition of σ:
$$\sigma^2 = \sum p(s)[X(s) - \mu]^2$$
By only summing some of the terms we cannot obtain a larger answer:
$$\sigma^2 \geq \sum_{|X(s) - \mu| \geq z\sigma} p(s)[X(s) - \mu]^2$$
By replacing $[X(s) - \mu]^2$ by its least possible value $z^2\sigma^2$, we similarly cannot increase the value of the sum:
$$\sigma^2 \geq \sum_{|X(s) - \mu| \geq z\sigma} p(s)z^2\sigma^2 = z^2\sigma^2 \sum_{|X(s) - \mu| \geq z\sigma} p(s)$$
Using Equation 6.26 we have
$$\sigma^2 \geq z^2\sigma^2(1 - p)$$
Dividing by $\sigma^2 z^2$, we have
$$\frac{1}{z^2} \geq 1 - p$$

Finally, adding $p - 1/z^2$ to both sides of this inequality, we obtain the result:

$$p \geqslant 1 - \frac{1}{z^2}$$

The numbers $1 - 1/z^2$ are called Chebyshev bounds. Some values are tabulated and graphed in Table 6.6.

6.6 Chebyshev Bounds

z	$1 - \dfrac{1}{z^2}$	z	$1 - \dfrac{1}{z^2}$
1	.000	3.16	.900
1.05	.100	4	.938
1.15	.250	4.47	.950
1.41	.500	5	.960
1.50	.556	7.07	.980
2	.750	10	.990
3	.889	31.62	.999

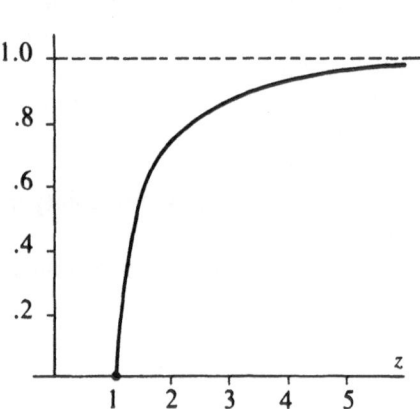

For example, taking $z = 1.05$, the value of X will be within 1.05σ of μ with probability at least .100. At the other extreme, Chebyshev's theorem shows that the value of X has probability at least 99.9 percent of falling within 31.62 standard deviations of μ.

9 Example

The average grade on an exam was 80 percent and the sample standard deviation was 5 percent. How many of the grades were between 65 and 95 percent? Within what range is it known that at least half the grades fell?

Method. We have $m = 80$, $s = 5$. Since $65 = 80 - 15 = m - 3s$, and $95 = 80 + 15 = m + 3s$, we are looking for the relative frequency that the grade X differed from the mean m by less than $3s$. Here $z = 3$, $1 - 1/z^2 = .889$. Thus more than 88 percent of the exams were in the range 65 through 95 percent.

For the second half we must find a value of z such that $1 - 1/z^2 = \frac{1}{2}$. Solving, we obtain $z = \sqrt{2} = 1.41$. (This can also be read from Table 6.6.) Thus at least half the grades fell within $1.42s$ of m. Since $1.42s = (1.42)(5) =$

7.1, we find that at least half the grades were in the range $80 - 7.1$ to $80 + 7.1$, or from 72.9 to 87.1.

We remark once again that Chebyshev's theorem yields a very conservative bound for the probability that $|X - \mu| < z\sigma$. For example, in Example 6 we found that fully 60 percent of the observations were within $z = 1$ standard deviation of the mean. Chebyshev's theorem guarantees 0 percent! Nonetheless, Chebyshev's theorem does give us information for $z > 1$. We can summarize as follows.

On the one hand, Chebyshev's theorem gives a universal estimate, valid for all random variables.[4] On the other hand, for a specific distribution, or for a special type of distribution, the estimate can be very weak.

EXERCISES

1. A random variable X has mean 8 and standard deviation .25. What can be said about the probability that X is between 7 and 9?

2. On some beautiful tropical island, the mean daily noon temperature during the year is $75°$ and the standard deviation is $3°$. An advertising agency wishes to use the statement "The noon temperature here is between $70°$ and $80°$ at least for x percent of the year." As a consultant you are required to find an honest value for x. Do so. Suppose the agency says that your answer is ridiculously conservative and demands a more realistic answer. How do you proceed?

3. In Exercise 2 the agency wants to use $x = 95$. Show how to do this by changing the extreme temperatures 70 and 80°.

4. A random variable X has $\mu = 3$ and $\sigma = 1$. What can you say about the probability that $X > 0$?

5. A random variable X has $\mu = 100$ and $\sigma = .1$. Find an interval of numbers such that X is in this interval with probability .99 or more.

6. For the random variable of Exercise 5, find an estimate for the probability that X is between 99.5 and 100.5.

7. For the random variable of Exercise 5, find an estimate for the probability that X is between 99.4 and 100.2.

*8. Theorem 8 shows that $p \geq 1 - 1/z^2$. By carefully going over the proof, prove (for fixed $z \geq 1$) that $p > 1 - 1/z^2$ except for one case: $X = \mu$, $\mu - z\sigma$, and $\mu + z\sigma$ with probability $1 - 1/z^2$, $1/2z^2$, and $1/2z^2$, respectively.

4 We are only considering random variables on finite sample spaces, although Chebyshev's theorem can be generalized to more complicated situations.

3 VARIANCE OF SUMS

In Section 4 of Chapter 5 we saw how the formula $E(X+Y) = E(X)+E(Y)$ was extremely useful for finding many expectations. It turns out that the analogous formula for $V(X)$ is valid for independent random variables.

10 Theorem
If X and Y are *independent* random variables on some sample space, then

$$V(X+Y) = V(X)+V(Y) \qquad (6.27)$$

Proof. Let $\mu_x = E(X)$ and $\mu_y = E(Y)$. By Theorem 46 of Chapter 5, $E(XY) = E(X)E(Y) = \mu_x\mu_y$. Also, $E(X+Y) = \mu_x+\mu_y$ by Theorem 27 of Chapter 5. Therefore, using Theorem 5, we have

$$\begin{aligned}
V(X+Y) &= E((X+Y)^2) - (\mu_x+\mu_y)^2 \\
&= E(X^2+2XY+Y^2) - (\mu_x^2+2\mu_x\mu_y+\mu_y^2) \\
&= E(X^2) + 2E(XY) + E(Y^2) - \mu_x^2 - 2\mu_x\mu_y - \mu_y^2 \\
&= E(X^2) + 2\mu_x\mu_y + E(Y^2) - \mu_x^2 - 2\mu_x\mu_y - \mu_y^2 \\
&= E(X^2) - \mu_x^2 + E(Y^2) - \mu_y^2 \\
&= V(X) + V(Y)
\end{aligned}$$

This completes the proof.

By an exactly similar method we can also prove the result for a sum of any finite number of independent random variables. We shall just state the result.

11 Theorem
Let $X_1, X_2, \ldots, X_k$ be *pairwise independent* random variables on some sample space. (This means that any two of these random variables are independent.) Then

$$V(X_1+\cdots+X_k) = V(X_1)+\cdots+V(X_k) \qquad (6.28)$$

Remark. It is the algebraic simplicity of these theorems which is ample justification for the use of $V(X)$ as a measure of dispersion.

If we translate Theorems 10 and 11 into statements about standard deviations, we obtain the following corollary:

12 Corollary
Let $X_1, \ldots, X_k$ be random variables, any two of which are independent. Let σ_i be the standard deviation of X_i. Let $X = \Sigma_i X_i$ and let $\sigma = \sigma(X) = \sigma(\Sigma X_i)$. Then

$$\sigma = \sqrt{\sum_i \sigma_i^2} \qquad (6.29)$$

In particular, if X and Y are independent with standard deviations σ_x and σ_y, then

$$\sigma(X+Y) = \sqrt{\sigma_x^2 + \sigma_y^2} \qquad (6.30)$$

The proof merely uses $\sigma^2 = V(X)$, $\sigma_i^2 = V(X_i)$. The result follows from Equation 6.28.

Remark. It must be remembered that Equations 6.27 through 6.30 are not true in general and that independence of any two of the random variables was assumed.

The value of the random variable X_i will be determined in many applications from the outcome of experiment i. If the experiments are independent, then the X_i are independent, by Theorem 51 of Chapter 3 and Definition 42 of Chapter 5, and the formulas may be used.

We now illustrate these theorems with some examples.

13 Example

Ten coins are tossed. Find the mean number of heads and the standard deviation.

Method. As in Example 29 of Chapter 5, we let N_i = the number of heads on coin i ($= 0$ or 1 each with probability $\frac{1}{2}$). Thus

$$N = N_1 + \cdots + N_{10} = \text{number of heads}$$

The N_k are independent. Therefore, setting $\sigma_k = \sigma(N_k)$, we have

$$\sigma(N) = \sqrt{\sigma_1^2 + \cdots + \sigma_{10}^2} \qquad (6.31)$$

Also,

$$E(N) = E(N_1) + \cdots + E(N_{10}) \qquad (6.32)$$

But each N_i has the same distribution $p(x)$ given by the formulas $p(0) = \frac{1}{2}$, $p(1) = \frac{1}{2}$. Therefore, $E(N_k) = \Sigma\, x_i p(x_i) = 0 \cdot \frac{1}{2} + 1 \cdot \frac{1}{2} = \frac{1}{2}$, and $V(N_k) = \Sigma\, (x_i - \mu)^2 p(x_i) = \frac{1}{4} \cdot \frac{1}{2} + \frac{1}{4} \cdot \frac{1}{2} = \frac{1}{4}$. Thus $\sigma_k^2 = \frac{1}{4}$. Finally, using Equations 6.31 and 6.32, we obtain

$$\sigma(N) = \sqrt{10 \cdot \tfrac{1}{4}} = \sqrt{2.5} \; (= 1.58)$$

and, as in Example 29 of Chapter 5,

$$E(N) = 10 \cdot \tfrac{1}{2} = 5$$

14 Example

Twenty dice are tossed. Find the average number of 6's tossed and the standard deviation for the number of 6's.

Method. Let N_k = number of 6's on the kth die. Then

$$N = \sum_{k=1}^{20} N_k = \text{number of 6's tossed}$$

The N_k are independent. Therefore, if we set $\mu_k = E(N_k)$ and $\sigma_k = \sigma(N_k)$, we have

$$\sigma(N) = \sqrt{\sum_{k=1}^{20} \sigma_k{}^2} \qquad E(N) = \sum_{k=1}^{20} \mu_k \qquad (6.33)$$

As in Example 13, each N_k has the same distribution. Here the distribution is given by Table 6.7, in which the computation of μ_k and σ_k is given.

6.7   Distribution for N_k

x	$p(x)$	$xp(x)$	$x^2p(x)$	
0	$\frac{5}{6}$	0	0	
1	$\frac{1}{6}$	$\frac{1}{6}$	$\frac{1}{6}$	$\sigma_k{}^2 = E(N_k{}^2) - \mu_k{}^2$
Sum		$\mu = \frac{1}{6}$	$E(N_k{}^2) = \frac{1}{6}$	$= \frac{1}{6} - \frac{1}{36} = \frac{5}{36}$

Thus $\mu_k = \frac{1}{6}$, $\sigma_k{}^2 = \frac{5}{36}$. Finally, from Equations 6.33, we find

$$\sigma(N) = \sqrt{20 \cdot \tfrac{5}{36}} = \tfrac{5}{3} \qquad E(N) = 20 \cdot \tfrac{1}{6} = \tfrac{10}{3}$$

15   Example

Six dice are tossed. Let A = average value of the numbers tossed. Find $E(A)$ and $\sigma(A)$.

Method.   If X_k = the number of the kth die, we have

$$A = \tfrac{1}{6}(X_1 + \cdots + X_6) \qquad (6.34)$$

Since the X_k are independent, we have

$$\sigma(A) = \sigma(\tfrac{1}{6}(X_1 + \cdots + X_6))$$
$$\sigma(A) = \tfrac{1}{6}\sigma(X_1 + \cdots + X_6)$$
$$\sigma(A) = \tfrac{1}{6}\sqrt{\sigma_1{}^2 + \cdots + \sigma_6{}^2} \qquad (6.35)$$

where $\sigma_k = \sigma(X_k)$. But $\sigma_1 = \sigma_2 = \cdots = \sigma_6$, because each X_k has the same distribution given by $p(x) = \frac{1}{6}$, $x = 1, 2, \ldots, 6$. It follows the $E(X_k) = \mu_k = \frac{1}{6}(1+2+\cdots+6) = \frac{7}{2}$ and $V(X_k) = \sigma_k{}^2 = \frac{1}{6}[(1-\frac{7}{2})^2 + (2-\frac{7}{2})^2 + \cdots + (6-\frac{7}{2})^2] = \frac{35}{12}$.

Thus, from Equation 6.35, $\sigma = \frac{1}{6}\sqrt{\frac{35}{12} \cdot 6} = \frac{1}{6}\sqrt{\frac{35}{2}} = .697$. From Equation 6.34,

$$E(A) = \tfrac{1}{6}[E(X_1) + \cdots + E(X_k)] = \tfrac{1}{6} \cdot 6 \cdot \tfrac{7}{2} = \tfrac{7}{2}$$

Thus A has $\mu = 3.5$, $\sigma = .697$. In contrast, X_1 has $\mu_1 = 3.5$, $\sigma_1 = \sqrt{\frac{35}{12}} = 1.71$.

The smaller standard deviation is plausible if we go back to the idea of experiments to determine A and X_1, respectively. *One* experiment for A is to toss 6 dice and average their sum. *One* experiment for X_1 is to toss 1 die and observe its value. It seems intuitively clear that the values obtained for A will not be as "spread out" as the values obtained for X.

EXERCISES

1. Two dice are tossed. Find the expected value and the standard deviation for the sum of the numbers that turn up.

2. Suppose that a printer makes an average of .1 errors per page. On the average, how many errors will appear in a 300-page book, and with what standard deviation? (Assume that the number of errors on each page is independent.)

3. A certain type of seed has probability .8 of germinating. One hundred seeds are planted. Find the expected number of germinations and the standard deviation.

4. Alan, Bob, and Carl choose integers at random. Alan chooses a number between 1 and 4 inclusive. Bob chooses a number between 1 and 10 inclusive and Carl chooses a number between 1 and 5 inclusive. Determine the expected value of the sum and its standard deviation.

5. A multiple choice test has 6 questions with a choice of 3 answers, 4 questions with a choice of 4 answers, and 10 true-or-false questions. Each question is worth 5 points. What is the mean grade if the answers are given at random? What is the standard deviation?

6. One hundred coins are tossed, and 60 of these coins land heads. How many standard deviations is this away from the expected number (50) of heads? (*Note:* More than 3 standard deviations is extremely unlikely.)

7. As in Exercise 6, 1,000 coins are tossed, and 600 land heads. How many standard deviations is this away from the expected number (500) of heads?

8. Politicians A and B are competing for the office of Chief Nay-sayer in an election. They both feel that everybody will vote at random and independently in this election, because nobody knows or cares about the politicians or the office. However, they know from past experience that people will vote for the *first name listed* with probability .51 and for the second name with probability .49. Since 1,000 people are going to vote, each feels that the first name is almost surely the winner. Taking for granted the numerical and political assumptions of A and B, how important is it to be listed first? Do the same if the election involves 100,000 people. (*Note:* A complete answer is not possible at this time, and must await Section 4 of Chapter 7. However, use Chebyshev's theorem for *some* estimate.)

9. If X and Y are independent random variables, prove that

$$V(aX + bY) = a^2 V(X) + b^2 V(Y)$$

Generalize to more than 2 variables.

10. By Equation 6.4, $V(2X) = 4V(X)$. But by Equation 6.26, $V(2X) = V(X + X) = V(X) + V(X) = 2V(X)$. Which is wrong, and why?

11. A gambling house has a game in which they lose \$10 with probability .1, win \$2 with probability .5, and win \$.50 with probability .4. On an average night, 1,000 independent games will be played. Compute the house's expected winnings and the standard deviation of winnings.

4 INDEPENDENT TRIALS

Throughout the course of the text we have made references to the notion of "repeating an experiment (independently) a large number of times." The very notion of probability was motivated in this way, as was the notion of the expected value of a random variable. We shall now put the idea of repeating an experiment independently into the formal framework of probability theory and relate the notions of mean and sample mean and of variance and sample variance.

In what follows we fix S as a finite sample space and let X be a random variable on S. It is desired to take N readings of X by repeating the experience for S independently N times. As we have seen in Chapter 3, the appropriate sample space to consider is $S^N = S \times S \times \cdots \times S$ (N factors). S^N consists of all N-tuples $(s_1, s_2, \ldots, s_N)$ with $s_i \in S$ and with $p(s_1, \ldots, s_N) = p(s_1) \cdots p(s_N)$. $(s_1, \ldots, s_N)$ may be regarded as a possible outcome of the sequence of N experiments (s_1 on the first experiment, etc.).

In what follows, we let X_i be the value of X determined on the ith experiment. Thus $X_i(s_1, \ldots, s_N) = X(s_i)$. The underlying sample space for X_1, $X_2, \ldots, X_N$ is S^N.

Using these ideas we are able to conceive of repeating an experiment N times and reading n values $x_1, \ldots, x_N$ of a random variable X as running *one* experiment in S^N and considering the N values of the N different random variables $X_1, \ldots, X_N$.

16 Theorem

The random variables $X_1, \ldots, X_N$ are independent. Each X_i has the same distribution as X.

Proof. Independence of the X_i means that the events A_i (in S^N) consisting of all s with $X_i(s) = x_i$ are independent for all choices of x_i. This is an immediate consequence of Theorem 52 of Chapter 3, because since A_i is determined by an event in the ith component of $S^N[X(s_i) = x_i]$. Also, if $p(x)$ is the distribution function of X, the probability that $X_i(s_1, \ldots, s_N) = x_k$ is the same as the probability that $X(s_i) = x_k$. (This follows from Equation 3.35.) Thus the distribution function of X_i is also $p(x)$. This completes the proof.

For example, if X is the number on a die and if $N = 4$, then we consider S^4, the sample space for 4 independent dice. X_3 is the number on the third die. Both X_3 and X have the same distribution given by $p(x) = \frac{1}{6}$ for $x = 1$, $2, \ldots, 6$.

17 Corollary

The mean and standard deviation of X_i are the same as the mean and standard deviation of X:

$$E(X_k) = E(X) = \mu \qquad \sigma(X_k) = \sigma(X) = \sigma \qquad V(X_k) = V(X)$$
$$(k = 1, \ldots, N) \qquad (6.36)$$

This is so because X_k has the same distribution as X.

The sample mean is simply the average of the N readings of X. We may therefore regard it as a random variable on S^N: Sum $X_1, \ldots, X_N$ and divide by N. The sample variance and sample standard deviation is similarly defined.

18 Definition

With $X, X_1, \ldots, X_N$ as above, we define[5]

$$\bar{X} = \frac{1}{N}(X_1 + \cdots + X_N) \tag{6.37}$$

$$S^2 = \frac{1}{N} \sum_{i=1}^{N} (X_i - \bar{X})^2 \qquad S = \sqrt{S^2} \tag{6.38}$$

Remark. It is appropriate that the sample mean and variance are random variables on S^N. For all this means is that the value of the sample mean (and variance) depends upon the outcomes $(s_1, \ldots, s_N)$ in S^N or, equivalently, upon the outcome of N experiments.

---

5 Some confusion in notation occurs here. We are using S to denote a sample space and S^N to denote the product of S with itself N times. Now we use S^2 to denote the sample variance and S the sample standard deviation. The usage will be clear from the context.

19 Theorem

With $\bar{X}$, μ, and σ defined as above, we have

$$E(\bar{X}) = \mu \qquad \sigma(\bar{X}) = \frac{1}{\sqrt{N}}\sigma \qquad V(\bar{X}) = \frac{1}{N}V(X) \qquad (6.39)$$

Proof. By Equations 6.36 and 6.37, and the properties of expectations, we have

$$E(\bar{X}) = E\left(\frac{1}{N}\sum_{i=1}^{N} X_i\right) = \frac{1}{N}\sum_{i=1}^{N} E(X_i) = \frac{1}{N}\sum_{i=1}^{N} \mu$$

$$= \frac{1}{N} \cdot N\mu = \mu$$

Since the X_i are independent, we may use Equation 6.28. Also, using Equation 6.4, we obtain

$$V(\bar{X}) = V\left(\frac{1}{N}\sum_{i=1}^{N} X_i\right) = \frac{1}{N^2}V\left(\sum_{i=1}^{N} X_i\right) = \frac{1}{N^2}\sum_{i=1}^{N} V(X_i)$$

$$= \frac{1}{N^2}\sum_{i=1}^{N} V(X) = \frac{1}{N^2}NV(X) = \frac{1}{N}V(X)$$

Finally, taking square roots, we obtain

$$\sigma(\bar{X}) = \frac{1}{\sqrt{N}}\sigma$$

This completes the proof.

Remark. We can paraphrase the result as follows: "If values of a random variable X (mean μ, standard deviation σ) are found in N independent experiments, then the mean value of the sample mean $\bar{X}$ is also μ, but the standard deviation of the sample mean $\bar{X}$ is $\sigma/\sqrt{N}$." As in the remark following Example 15, we expect the values of the average $\bar{X}$ to be less dispersed than the values of X.

We might suppose that the expected value of S^2 is σ^2. It is not.

*20 Theorem

If S^2 is the sample variance and σ^2 the variance of a random variable, we have

$$\sigma^2 = \frac{N}{N-1}E(S^2) \qquad (6.40)$$

Proof. Using Equations 6.38 and 6.25, we have

$$S^2 = \frac{1}{N}\sum_{i=1}^{N} (X_i - \bar{X})^2 = \frac{1}{N}\left(\sum_{i=1}^{N} X_i^2\right) - \bar{X}^2$$

We take expected values, using Equation 6.16 $[E(X^2) = \sigma^2 + \mu^2]$. Thus

$$E(S^2) = E\left(\frac{1}{N}\sum_{i=1}^{N} X_i^2\right) - E(\bar{X}^2)$$

$$= \frac{1}{N}\sum_{i=1}^{N} E(X_i^2) - E(\bar{X}^2)$$

$$= \frac{1}{N}\sum_{i=1}^{N} (\sigma^2 + \mu^2) - \left(\frac{1}{N}\sigma^2 + \mu^2\right)$$

$$= \sigma^2 + \mu^2 - \left(\frac{\sigma^2}{N} + \mu^2\right)$$

$$= \frac{N-1}{N}\sigma^2$$

After multiplying by $N/N-1$, we obtain the result.

Remark. σ^2 is a *number* whose value depends on how the values of X are distributed on the underlying sample space. S^2 is a *random variable* whose value depends on a point in S^N and on the random variable X (i.e., on the outcome of N experiments, and the value of X on each experiment). The theorem relates σ^2 and S^2: $\sigma^2 = E((N/N-1)S^2)$. If you do not know the mean of a random variable X and if you are allowed one experiment, yielding outcome s_0 and value $x_0 = X(s_0)$, then x_0 is a good guess for μ. (The smaller σ is, the more likely it is that x_0 will be close to μ.) When we compute the sample variance s^2 for a series of experiments, we are, in fact, finding one value of S^2 as the result of one experiment (in S^N). Since $E((N/N-1)S^2) = \sigma^2$, we consider $(N/N-1)s^2$ to be a good estimate for σ^2. But

$$\frac{N}{N-1}s^2 = \frac{N}{N-1}\frac{1}{N}\sum_{i=1}^{N} (x_i - \bar{x})^2 = \frac{1}{N-1}\sum_{i=1}^{N} (x_i - \bar{x})^2$$

For this reason many authors prefer using

$$\frac{1}{N-1}\sum_{i=1}^{N} (x_i - \bar{x})^2$$

as their definition of sample variance. It is a more sensible estimate for σ^2.

21 Example
A biased coin has probability .4 of landing heads and probability .6 of landing tails. Fifty independent tosses are used to determine the relative frequency F of heads. What is the expected value of F and the standard deviation of F?

Method. Let $X = 1$ if the coin lands H and let $X = 0$ if the coin lands T. The relative frequency F is simply

$$F = \tfrac{1}{50} \times (\text{number of heads in 50 trials})$$
$$= \tfrac{1}{50}(X_1 + X_2 + \cdots + X_{50})$$
$$= \bar{X}$$

Thus, by Equation 6.39,

$$E(F) = \mu \qquad (F) = \frac{\sigma}{\sqrt{50}}$$

where $E(X) = \mu$, $\sigma(X) = \sigma$. But $E(X)$ and $\sigma(X)$ can be easily computed as in the following table from the distribution of X:

x	$p(x)$	$xp(x)$	$x^2 p(x)$
0	.6	0	0
1	.4	.4	.4
Sum		$\mu = .4$	.4: $\qquad \sigma^2 = .4 - (.4)^2 = .24, \sigma = \sqrt{.24}$

Thus $\mu = .4$, $\sigma = \sqrt{.24}$, and finally we obtain

$$E(F) = .4 \qquad \sigma(F) = \frac{\sqrt{.24}}{\sqrt{50}} = .069$$

Thus we not only obtain $E(F) = .4$, which we might have anticipated, but we also obtain the value $\sigma(F) = .069$, which is a measure of how far the observed value of the relative frequency will be from .4. For example, using the very weak Chebyshev inequality for $z = 2$, we find that $|F - .4| < .138$ with probability greater than .75.

EXERCISES

1. One hundred people independently shuffle a pack of cards and observe whether the top card is a spade or not. The relative frequency of spades is then computed. What is the mean relative frequency? What is the standard deviation of the relative frequency? [*Hint:* Let $X(s) = 1$ if s is a spade, $X(s) = 0$ if s is not a spade. Then $\bar{X}$ is the relative frequency.]

2. It is desired to have a large number of people perform the experiment of Exercise 1. It is also desired that the relative frequency of spades should almost certainly be between .2 and .3. How many people should we have perform the experiment? (*Hint:* Take "almost certainly" to mean probability .9, .95, and .99 to obtain 3 answers. Use Chebyshev's theorem for very conservative answers. In Chapter 7 we shall be less conservative.)

*3. We found that $E((N/N-1)S^2) = \sigma^2$. Explain why $V(S^2)$ or $\sigma(S^2)$ cannot be found in terms of μ and σ.

4. Find $E((1/N) \sum_{i=1}^{N} (X_i - \mu)^2)$.

5. A random variable X has mean 50 and standard deviation 2. How many experiments should be performed so that $\bar{X}$ will have standard deviation .4? Using Chebyshev's theorem, how many experiments should be performed so that $\bar{X}$ is between 49 and 51 with probability at least .9?

6. Suppose $N = 3$ independent trials occur. Define $M_1 = \frac{1}{6}(3X_1 + 2X_2 + X_3)$. Find $E(M_1)$ and $V(M_1)$. Explain why $\bar{x} = \frac{1}{3}(x_1 + x_2 + x_3)$ is, other things equal, a better estimate for the mean μ than $m_1 = \frac{1}{6}(3x_1 + 2x_2 + x_3)$. (The number m_1 is called a weighted mean.)

*7. Generalize Exercise 6 as follows: If $M_1 = \sum_{i=1}^{N} a_i X_i$, where $\sum a_i = 1$, then the choice of a_i that minimizes $\sigma(M_1)$ is $a_i = 1/N$. [*Hint:* First find $V(M_1)$ and then set $a_i = (1/N) + b_i$.]

8. During 5 independent trials, the values of a random variable X were observed to be $x = 5, 5.2, 5.3, 4.7$, and 5. Estimate the mean and variance of X.

*9. What can you say about the value of $E(S)$? If a sample variance s is found, how can you use it to estimate σ? (*Hint:* Exercise 6, Section 6.1.)

10. A random variable X has $\mu = 10$ and $\sigma = .2$. Ten independent samples of X will be taken and the sample average $\bar{X}$ computed. Give an estimate for the probability that $|\bar{X} - 10| < .4$.

11a. An experiment consists of tossing 2 coins and recording the number H of heads. (H = 0, 1, or 2). Find $\mu = \mu(H)$ and $\sigma = \sigma(H)$.
   b. The experiment is repeated $N = 10$ times. Find $\mu(\bar{H})$ and $\sigma(\bar{H})$.
   c. Perform this experiment 10 times and find the sample mean m and the sample variance s^2 for your results. Compare m and μ. Compute $(N/N-1)s^2 = \frac{10}{9}s^2$ and see how it compares to σ^2.

5 THE LAW OF LARGE NUMBERS

We now have enough apparatus to consider what happens if the number N of experiments is made very large. The following theorem shows how the sample mean of a random variable is related to the mean if the number of trials is large.

22 Theorem. Law of Large Numbers

Let X be a random variable on a finite sample space having mean μ and standard deviation σ. Let $d > 0$ be given. Let $p_N =$ probability that $\mu - d < \bar{X} < \mu + d$. Then $p_N \to 1$ as $N \to \infty$.

The limit statement can be paraphrased as follows: Let $d, f > 0$ be given. Then some N_0 may be determined so that if $N \geq N_0$, the probability that $\mu - d < \bar{X} < \mu + d$ is at least $1 - f$.[6] It is this latter statement that we shall use and prove.

Remark. We can thus say that the sample mean s will very likely be very close to μ if a large number of trials are taken. The "very likely" is measured by f, the "very close" is measured by d, and the "large" is measured by N_0. If, for example, $\mu = .3$, $d = .02$, $f = .001$, then we can find N_0 so that $.28 < \bar{X} < .32$ with probability at least $.999$. Also, if more than N trials are taken, then $.28 < \bar{X} < .32$ also with probability at least $.999$. (We have a different $\bar{X}$, because we choose a different N.)

Proof. We note that the equation $\mu - d < \bar{X} < \mu + d$ means the same as $|\bar{X} - \mu| < d$. According to Chebyshev's theorem (Theorem 7), the probability that $|\bar{X} - \mu| < z\sigma/\sqrt{N}$ is larger than $1 - (1/z^2)$. ($\sigma/\sqrt{N}$ is used, because $\sigma/\sqrt{N}$ is the standard deviation of $\bar{X}$. We are applying Chebyshev's theorem to $\bar{X}$.) We choose z so that

$$d = \frac{z\sigma}{\sqrt{N}}$$

Solving, we have

$$z = \frac{d\sqrt{N}}{\sigma}$$

Thus

$$1 - \frac{1}{z^2} = 1 - \frac{\sigma^2}{d^2 N}$$

We can arrange $1 - 1/z^2$ to be at least $1 - f$ by arranging to have

$$1 - \frac{\sigma^2}{d^2 N} \geq 1 - f$$

or

$$f \geq \frac{\sigma^2}{d^2 N}$$

Multiplying by N/f, this is equivalent to

$$N \geq \frac{\sigma^2}{d^2 f} \qquad (6.41)$$

---

6 Note that N enters the condition $\mu - d < \bar{X} < \mu + d$, because $\bar{X}$ is defined in terms of N. The notation $\bar{X}$ does not make this dependence clear.

If this inequality holds, then the required probability will be larger than $1 - (1/z^2)$, hence larger than $1 - f$. This completes the proof.

Note that the estimate is based on Chebyshev's theorem and hence is very conservative. N_0 is taken as any integer greater or equal to $\sigma^2/d^2 f$. For the example given in the above remark, suppose that $\sigma = .1$. Then Equation 6.35 prescribes

$$N \geq \frac{.01}{(.001)(.0004)} = 25,000$$

Thus the conservative estimate says that if 25,000 or more trials are taken, the sample mean will be in the interval $.28 < \bar{X} < .32$ with probability 99.9 percent. (Actually, a more realistic answer is $N = 275$.) As we can see by the inequality 6.41, the number N of samples will be increased if σ is increased or if d or f is decreased. This is the price we must pay to get closer to the mean and for a higher probability of attaining that accuracy.

The above theorem relates the sample mean for large samples to the mean. We can now relate the relative frequency of an event to the probability of that event. (Cf. Chapter 1!) The trick is to observe, as in Example 21, that the relative frequency is the sample average of a particularly simple random variable.

23 Theorem. Bernoulli's Theorem

Let S be a finite sample space and let $A \subseteq S$ be an event of S with probability $p(A) = p$. Let $F = F_N$ be the relative frequency of the occurrence of A in N trials. (F is a random variable on S^N.) Let $d > 0$ be given and let $p_N = $ probability that $|F(s) - p| < d$. Then $p_N \to 1$ as $N \to \infty$.

As in Theorem 21, this theorem can be stated without the use of limits. Thus, *if d and f are given positive numbers, there is an N_0 such that if $N \geq N_0$, the probability that $|F_N(s) - p| < d$ is at least $1 - f$.* This is the form in which the theorem will be proved.

We may loosely say that the relative frequency of A will very likely be very close to A provided a large number of experiments are performed.

Proof. Define the random variable Y on S as follows:

$$Y(s) = \begin{cases} 1 & \text{for } s \text{ in } A \\ 0 & \text{for } s \text{ not in } A \end{cases}$$

Then $\bar{Y}$ is simply the relative frequency:

$$\bar{Y}(s) = F_N(s) \qquad s \in S^N$$

In fact, $\bar{Y}(s) = (1/N)[Y_1(s) + \cdots + Y_N(s)] = (1/N)[Y(s_1) + Y(s_2) + \cdots + Y(s_N)]$ is the number of times s_i is in A, divided by N. The distribution of $p(x)$ is given by the equations $p(1) = p$, $p(0) = 1 - p$. Therefore, $E(Y)$ is simply $\Sigma xp(x) = 1 \cdot p + 0 \cdot (1 - p)$. Thus, by Theorem 22, we can find N_0

so that

$$|\bar{Y}-p| = |F_N-p| < d$$

with probability at least $1-f$, for all $N \geqslant N_0$. This completes the proof.

We conclude this chapter with two remarks. First, Theorem 22 has been greatly generalized to infinite probability spaces, so it must be regarded as a relatively weak form of the law of large numbers. Second, it must not be thought that Theorem 23 justifies the notion of statistical probability as a limit of relative frequencies, for that theorem makes a statement that some inequality holds with probability close to 1. But it is still framed in the language of probability theory and so can hardly be used to justify probability as a limit of relative frequencies. There are stronger statements possible, but the fact remains that one could take a fair coin and toss it once a day for 10 years and it *might* land heads each time. No theory of probability has changed that possibility.

EXERCISES

1. Compute $\sigma(Y)$ in the proof of Theorem 23, and in this way find the number N_0 of experiments necessary to guarantee that the relative frequency of A will be within d of the probability $p(A)$, with probability greater than $1-f$.

2. In Exercise 1 find a number N_1 that is independent of p and depends only on d and f. [*Hint:* $p-p^2 = \frac{1}{4} - (\frac{1}{2}-p)^2$.]

3. You have a die (possibly loaded), and you want to demonstrate, experimentally, that if it is tossed independently for a large number of times the relative frequency will have some limit. Therefore, you announce to the class: "I am going to toss this die N times and compute the relative frequency of the event 1 or 6." Tomorrow, I will again toss it N times and compute the same relative frequency. I predict that the two relative frequencies will be within .01 of each other." As a teacher you are willing to have your prediction fail, but you want at least 90 percent chance of success. What number N do you announce? (Use Chebyshev's estimate, and try to make both relative frequencies within .005 of the mean.)

4. In Exercise 3, let F_1 be the relative frequency on the first day and let F_2 be the relative frequency on the second day. Let $E = F_2 - F_1$. Using Exercise 9 of Section 3 ($a = 1$, $b = -1$), find N so that $|E| < .01$. Explain why this N is smaller than the N found in Exercise 3.

CHAPTER 7 SOME STANDARD DISTRIBUTIONS

1 CONTINUOUS DISTRIBUTIONS

So far, the distributions $p(x)$ that we have considered were defined for finitely many values $x_1, \ldots, x_n$. We shall now consider distributions $f(x)$ defined for *all* real numbers x or for real numbers x in some interval. This is a big step to take, so it is worth pausing to ask why we do so.

One could argue that this step is not really necessary. After all, laboratory experiments effectively involve only a finite number of possibilities. Furthermore, even if there is a continuum of possibilities (say, in determining the length of a piece of wire), our measurements are only good to so many decimal places. It would appear to be unnecessary to consider all possibilities. For example, if a piece of wire is to be measured in a laboratory, if it is known that it is about 10 cm long, and if measurements are accurate to 2 decimal places, then the variability in the measurement *could* be expressed by a distribution $p(x)$, where $x = 8.00,\ 8.01, \ldots,\ 10.00, \ldots,$ 11.99, 12.00. However, it turns out that it is often even *simpler* to consider a continuous distribution $p(x)$, where x is a real variable.

Continuous and discrete mathematics each throw light on the other. For example, modern high-speed (digital) computers always reduce continuous problems to discrete ones. On the other hand, many discrete physical situations are greatly simplified by working continuously. A chemist, for example, will analyze the *mass* of a certain amount of radium rather than the *number* of atoms present.

1 Example

A real number x is chosen uniformly and at random between 0 and 3. What is the probability that $1 \leq x \leq 2$? Similarly, find the probability that $2\frac{1}{2} \leq x \leq 5$.

Method. The given interval $0 \leq x \leq 3$ has length 3, while the interval $1 \leq x \leq 2$ has length 1. Intuitively we would expect the probability that x is in any interval to be proportional to the length of that interval. Therefore, we expect the probability that $1 \leq x \leq 2$ [written $p(1 \leq x \leq 2)$] to be $\frac{1}{3}$. In general, if a and b are in the given interval from 0 to 3, then $p(a \leq x \leq b) = \frac{1}{3}(b-a)$, because the interval $a \leq x \leq b$ has length $b-a$, and the original interval from 0 to 3 has length 3. To find $p(2\frac{1}{2} \leq x \leq 5)$, we note that x cannot be larger than 3. Thus $p(2\frac{1}{2} \leq x \leq 5) = p(2\frac{1}{2} \leq x \leq 3)$. Since $3 - 2\frac{1}{2} = \frac{1}{2}$, we have $p(2\frac{1}{2} \leq x \leq 5) = (\frac{1}{3})(\frac{1}{2}) = \frac{1}{6}$.

There is a very useful geometric way of picturing the probability $p(a \leq x \leq b) = \frac{1}{3}(b-a)$. In Fig. 7.1 we draw the graph of $y = \frac{1}{3}$ for $0 \leq x \leq 3$. Then the value $b-a$ is the length of the interval from a to b, and the probability $\frac{1}{3}(b-a)$ is the *area* under the graph $y = \frac{1}{3}$ and between the values $x = a$ and $x = b$. If we define $y = 0$ for $x < 0$ and for $x > 3$ (the darkened lines in Fig. 7.1), the area also gives the correct formula for any a and b. Thus the area under the graph from $x = 2\frac{1}{2}$ to $x = 5$ is $(\frac{1}{2})(\frac{1}{3}) = \frac{1}{6}$, because there is zero-area contribution from $x = 3$ to $x = 5$.

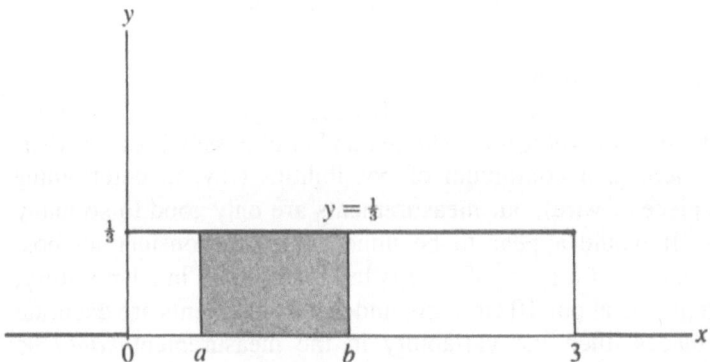

7.1 *Distribution for a Uniform Random Variable Between 0 and 3*

For this example we define $f(x)$ as follows:

$$f(x) = \begin{cases} \frac{1}{3} & \text{for } 0 \leq x \leq 3 \\ 0 & \text{for } x < 0 \text{ and for } x > 3 \end{cases}$$

Then, when x is chosen uniformly between 0 and 3, the probability $p(a \leq x \leq b)$ is given as the area between the graph of $y = f(x)$ and the x

axis, which is bounded on the left by $x = a$ and on the right by $x = b$. Note that the total area under the graph is 1. Thus x falls somewhere with probability 1.

For finite distributions $p(x)$ $(x = x_1, \ldots, x_n)$, we have two fundamental properties:

$$p(x_i) \geq 0 \qquad (i = 1, \ldots, n) \tag{7.1}$$

$$\sum_{i=1}^{n} p(x_i) = 1 \tag{7.2}$$

We note that for the function $f(x)$ defined above, we had $f(x) \geq 0$, in analogy with condition 7.1. The area under the graph $y = f(x)$ was 1. This is in analogy with condition 7.2.

We can generalize the above example greatly to arbitrary continuous distributions. We first introduce a notation for area.

2 Definition

Let $y = f(x)$ be a function of x such that $f(x) \geq 0$. If $a < b$, we define $A_f(a, b)$ to be the area between the graph $y = f(x)$, the x axis, to the right of $x = a$ and to the left of $x = b$.[1] (See Fig. 7.2.)

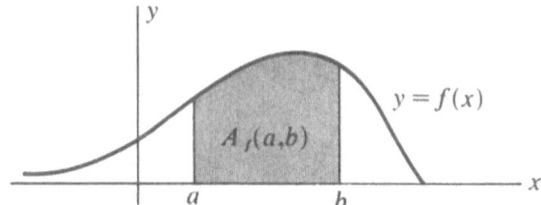

7.2   Area Under a Curve

In this text, areas under curves will be found directly from tables, unless (as in Example 1) these areas can be easily computed using elementary geometry. We now generalize Example 1 to arbitrary continuous distributions.

3 Definition

A *continuous probability density* is a function $f(x)$ such that

$$f(x) \geq 0 \tag{7.3}$$

$$A_f(-\infty, +\infty) = 1 \tag{7.4}$$ [2]

---

1 In calculus notation this area is written $\int_a^b f(x)\, dx$.

2 Thus the entire area under the curve $y = f(x)$ is 1.

We say that the variable x has the probability density $f(x)$ if

$$p(a \leqslant x \leqslant b) = A_f(a, b) \tag{7.5}$$

for all choices of a and b.

In Example 1 the variable x had the probability density $f(x)$, where $f(x) = \frac{1}{3}$ for $0 \leqslant x \leqslant 3$ and $f(x) = 0$ for $x < 0$ and for $x > 3$. Sometimes $f(x)$ is only defined in an interval $a_0 \leqslant x \leqslant b_0$. In that case we can always define $f(x) = 0$ outside that interval so that $A_f(-\infty, +\infty)$ has a meaning for every f. Equation 7.4 then becomes $A_f(a_0, b_0) = 1$, because there is no contribution of area outside the interval.

4 Example

A die is tossed. If it lands 5 or 6, a number x is chosen uniformly at random in the interval 0 to 3. If it lands 1, 2, 3, or 4, a number x is chosen uniformly at random in the interval 3 to 4. Find the density function for x. Use it to find $p(2 \leqslant x \leqslant 3.6)$.

Method. The density function is as in Fig. 7.3, because x is uniformly distributed in $0 \leqslant x \leqslant 3$ and in $3 \leqslant x \leqslant 4$. Since $p(0 \leqslant x \leqslant 3) = \frac{1}{3}$ (the probability that the die lands 5 or 6), we have $3c = \frac{1}{3}$ and $c = \frac{1}{9}$. Similarly, $1d = d = \frac{2}{3}$. Thus $f(x) = \frac{1}{9}$ for $0 \leqslant x \leqslant 3$, $f(x) = \frac{2}{3}$ for $3 < x \leqslant 4$. Thus $p(2 \leqslant x \leqslant 3.6) = A_f(2, 3.6) = \frac{1}{9} + \frac{2}{3}(.6) = .111 + .4 = .511$.

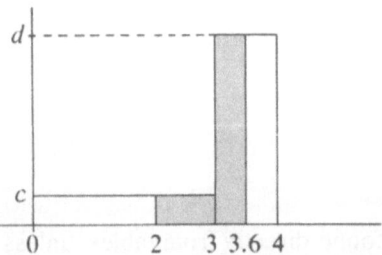

7.3

5 Example

The probability distribution $f(x)$ is given by the formulas

$$f(x) = \begin{cases} x & 0 \leqslant x \leqslant 1 \\ 2-x & 1 \leqslant x \leqslant 2 \end{cases}$$

Find $p(\frac{1}{2} \leqslant x \leqslant \frac{5}{3})$.

Method. The graph consists of the 2 line segments of Fig. 7.4. Note that the total area under the "curve" $y = f(x)$ is 1. It is required to compute the shaded area. We use the formula $K = \frac{1}{2}h(b_1 + b_2)$ for the area for a trapezoid to obtain

$$A = \left(\tfrac{1}{2}\right)\left(\tfrac{1}{2}\right)\left(1+\tfrac{1}{2}\right) = \tfrac{3}{8}$$

$$B = \left(\tfrac{1}{2}\right)\left(\tfrac{2}{3}\right)\left(1+\tfrac{1}{3}\right) = \tfrac{4}{9}$$

The required probability is thus $A\left(\tfrac{1}{2},\tfrac{5}{3}\right) = A + B = \tfrac{3}{8} + \tfrac{4}{9} = \tfrac{59}{72}$ $(= .82)$.

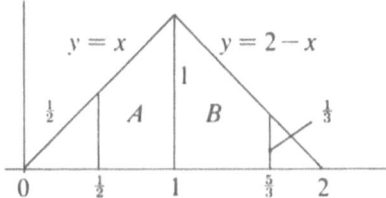

7.4   Distribution

6   Example

An integer N from 1 through 3,000 is chosen at random. Find the probability that N is in the range 200 through 2,200.

Method.   We can compute $p(200 \leqslant N \leqslant 2{,}200)$ explicitly. Each integer has probability $1/3{,}000$. There are $2{,}200 - 200 + 1 = 2{,}001$ integers in the given range. Thus the probability is $2{,}001/3{,}000 = .667$ (exactly).

   We can approximate this problem continuously. Let $x = N/1{,}000$. Then $x = .001, .002, \ldots, 3.000$, with equal probabilities. We therefore think of choosing a continuous variable x uniformly in the range $0 \leqslant x \leqslant 3$, leading to the distribution of Example 1. The condition $200 \leqslant N \leqslant 2{,}200$ is, after dividing by 1,000 and setting $N/1{,}000 = x$, equivalent to $.2 \leqslant x \leqslant 2.2$. We are thus led to Fig. 7.5, where $A(.2, 2.2)$ is seen to be $(2.0)\tfrac{1}{3} = \tfrac{2}{3} = .667$ (to

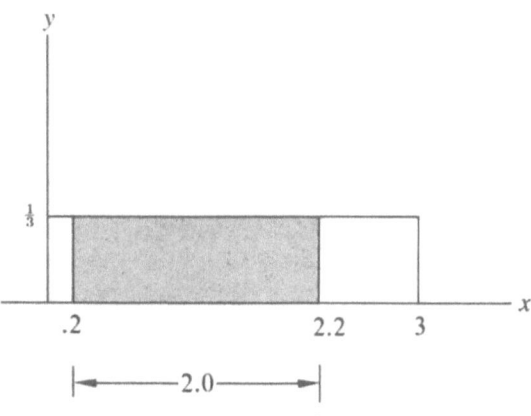

7.5

3 decimal places). Thus the continuous approximation of the finite distribution yields the correct answer to 3 decimal places and gives a rather nice picture of what is happening.

Two features of this simple example will carry over to later, more complicated, situations. *First, the change of scale* in which N was converted into $x = N/1,000$ served to compress the values of N closer together so that a continuous approximation was feasible. Figures 7.6a, 7.6b, and 7.6c show the distributions for N, $N/1,000$, and x. (The first 2 are discrete, and the last is continuous.) The difference between Figs. 7.6b and 7.6c is very pronounced. Note that the heights 1/3,000 barely shows up in Fig. 7.6b. (These heights are actually exaggerated there.) The height $\frac{1}{3}$ in Fig. 7.6c is more tangible. Why is this? In Fig. 7.6b the way to compute the required probability is to *sum the many heights* $p(x)$ at $x = .200, .201, \ldots, 2.200$. But in Fig. 7.6c the probability was found as an area. For a probability density $f(x)$ it is the area, not the height, which is a probability.

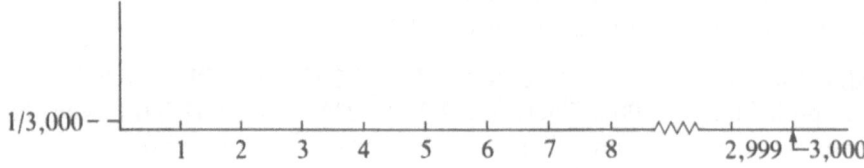

7.6a   Distribution for N

7.6b   Distribution for N/1,000

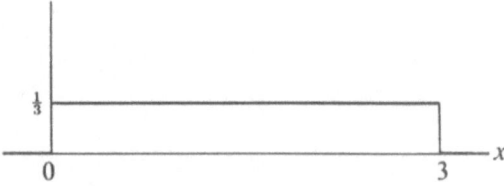

7.6c   Probability Density for x

The second feature of this example that will carry over to other examples is the *indeterminacy of the end points for the discrete variable N*. For example, instead of writing $220 \leqslant N \leqslant 2{,}200$, we could just as well write $219 < N < 2{,}201$, or $219.7 \leqslant N \leqslant 2{,}200.8$. All these inequalities yield $N = 220, 221, \ldots, 2{,}200$, because N is an integer. Since the lower bound of N can be as large as 220 (inclusive) or as small as 219 (exclusive), a good approximation will be obtained if the average 219.5 is used. Similarly, 2,200.5 is also used. The inequality $219.5 \leqslant N \leqslant 2{,}200.5$, after dividing by 1,000, leads to $.2195 \leqslant x \leqslant 2.2005$. In this case $A(.2195, 2.2005)$ will actually yield the *exact* answer for the probability that $220 \leqslant N \leqslant 2{,}200$. The probability that $N = 500$ (for example) can be found similarly. Merely write $N = 500$ as *the inequality* $499.5 \leqslant N \leqslant 500.5$ and divide by 1,000 to obtain the equivalent inequality $.4995 \leqslant x \leqslant .5005$, whose probability is $A(.4995, .5005) = (\tfrac{1}{3})(.0010) = 1/3{,}000$.

EXERCISES

1. A probability density is given by the equation $f(x) = \tfrac{1}{4}$ for $-2 \leqslant x \leqslant 2$. Find

    a. $p(1.3 \leqslant x \leqslant 1.9)$ **b.** $p(-1.2 \leqslant x \leqslant 1.1)$
    c. $p(1.3 \leqslant x \leqslant 7)$ **d.** $p(-2.6 \leqslant x \leqslant -1.3)$

2. A probability density is given by the equations $f(x) = \tfrac{1}{2}$ for $0 \leqslant x \leqslant 1$ and $f(x) = \tfrac{1}{4}$ for $1 < x \leqslant 3$. Graph $y = f(x)$. Find $p(\tfrac{1}{2} \leqslant x \leqslant \tfrac{3}{2})$. Find $A_f(0, 2)$ and $A_f(.5, 2.5)$.

3. If $f(x)$ is any probability density and $a < b < c$, explain why $A_f(a, c) = A_f(a, b) + A_f(b, c)$. What is the probabilistic significance of this equation? What is the analogue for a finite distribution $p(x)(x = x_1, \ldots, x_n)$?

4. The function $y = kx$ is a probability density in the interval $0 \leqslant x \leqslant 4$. Find k.

5. A die is tossed. If it lands 3 or more, a number x is chosen at random between 0 and 1. If it lands 1 or 2, x is chosen at random between 1 and 2. Find the probability density for x. Find the probability that $\tfrac{1}{2} \leqslant x \leqslant 1\tfrac{1}{4}$.

6. As in Exercise 5, suppose that x is chosen at random between 0 and 2 if the die lands 3 or more, and that x is chosen at random between 1 and 3 if the die lands 1 or 2. Find the probability density for x. Using this density, find $p(\tfrac{1}{2} \leqslant x \leqslant 2)$. Similarly, find $p(1\tfrac{1}{2} \leqslant x \leqslant 2\tfrac{1}{2})$.

2 NORMAL DISTRIBUTION

One of the distributions that frequently arises in applications is the normal distribution. One important application of this continuous distribution is to approximate a discrete distribution.

7 Definition

The *standardized normal probability density function* is the function $n(z)$ given by the formula

$$n(z) = \frac{1}{\sqrt{2\pi}} e^{-z^2/2} \tag{7.6}$$

The normal curve $y = n(z)$ is the familiar bell-shaped curve so well loved by psychologists, mathematicians, biologists, and educators. Its graph is given in Fig. 7.7, along with some typical values for y. Here $\pi = 3.14159\ldots$, of geometry fame, while $e = 2.71828\ldots$ is the number introduced in Chapter 4. We shall *never* again be concerned with Equation 7.6. Its values have been extensively computed and tabulated, as have the areas $A_n(a, b)$. We shall shortly learn how to find these areas from tables.

7.7 Standardized Normal Curve

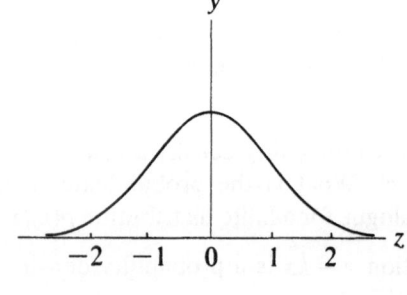

$\pm z$	y	$\pm z$	y
0.0	.40	2.5	.02
0.5	.35	3.0	.0044
1.0	.24	3.5	.0009
1.5	.13	4.0	.0001
2.0	.05		

(y to 2 decimal places, except for larger values of z.)

As we can see from Fig. 7.7, the normal curve is symmetric about the y axis, and y rapidly approaches 0 as z approaches $\pm\infty$. In fact, even $z = 4$ gives $y = .0001$, and when $z = 5$, $y = .0000015$. Even though the curve extends to infinity, it dies down so rapidly that for most purposes we need only consider values z between -3 and 3. Before using the normal distribution in what follows, it will be necessary to take for granted several facts about it. We state them here.

1. The area under the curve $y = n(z)$ is 1. $[A_n(-\infty, +\infty) = 1.]$ This is difficult to prove and is a very deep result.

2. The curve $y = n(z)$ is symmetric about the y axis. Thus computations involving $n(z)$ may be performed for $z \geq 0$, since the graph for $z \leq 0$ is the mirror image, about the y axis of the curve for $z \geq 0$. In particular, $A_n(0, \infty) = .5$.

3. The values of the areas $A_n(0, z)$ have been computed and can be obtained from Appendix D. We write $A_n(0, z)$ briefly as $N(z)$. A brief summary is given in Fig. 7.8.

7.8 Areas Under the Standardized Normal Curve

z	$N(z)$	z	$N(z)$	z	$N(z)$
0.0	.00	0.9	.32	1.8	.46
0.1	.04	1.0	.34	1.9	.47
0.2	.08	1.1	.36	2.0	.48
0.3	.12	1.2	.38	2.1	.48
0.4	.16	1.3	.40	2.2	.49
0.5	.19	1.4	.42	2.3	.49
0.6	.23	1.5	.43	2.4	.49
0.7	.26	1.6	.45	2.5	.49
0.8	.29	1.7	.46	2.6	.50

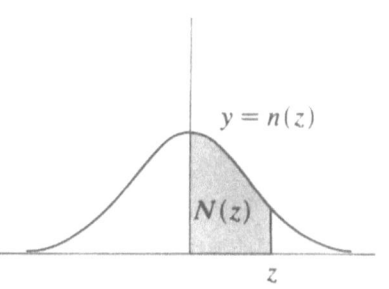

4. The normal distribution has mean 0 and standard deviation 1. In this text we have not even defined the mean and standard deviation of a continuous distribution, so this statement is, within the context of this text, meaningless. However, a definition can be given for continuous distributions, in which the normal distribution turns out to have mean 0 and standard deviation 1.

We can now define a normal distribution with arbitrary mean μ and standard deviation σ.

8 Definition

Suppose the variable x has a probability distribution $f(x)$. Suppose also that μ and $\sigma > 0$ are two numbers such that, for all $a < b$,

$$p(\mu + a\sigma \leqslant x \leqslant \mu + b\sigma) = A_n(a, b) \tag{7.7}$$

where $A_n(a, b)$ is the area function for the standardized normal distribution. We then say that x is *normally distributed with mean μ and standard deviation σ.*

Another way of looking at this definition is as follows. Suppose we set

$$x = \mu + z\sigma \tag{7.8}$$

Then the condition that $\mu + a\sigma \leqslant x \leqslant \mu + b\sigma$ is equivalent to $a \leqslant z \leqslant b$. The variable z, already used in the section on Chebyshev's theorem, measures how many standard deviations x is away from the mean. Definition 8 can therefore be stated as follows: x is normally distributed with mean μ and standard deviation $\sigma > 0$ provided z has the standardized normal probability density.

The relationship between x and z, and the probabilities $p(\mu + a\sigma \leqslant x \leqslant \mu + b\sigma)$ and $p(a \leqslant z \leqslant b)$, are illustrated in Fig. 7.9. The standardized normal curve $y = n(z)$ is shifted over and is centered over at μ. The z axis is then magnified by an amount σ, and the vertical coordinates are compressed by a like amount (thereby keeping areas the same). The figure illustrates $p(\mu - \sigma \leqslant x \leqslant \mu + 2\sigma) = A_n(-1, 2)$.

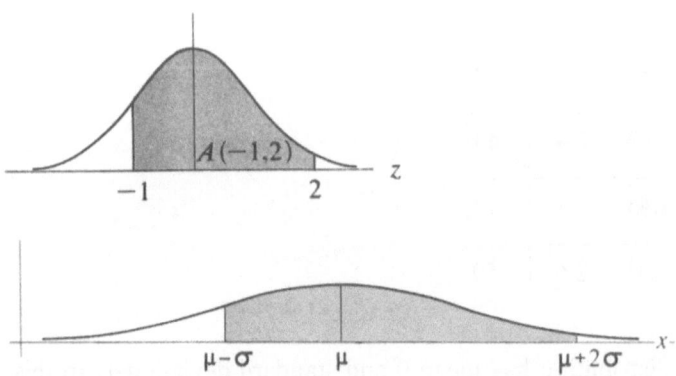

7.9 Relationship Between Normal Variable x and Standardized Variable z

As a special case of Definition 8, we note that a variable z which has the standardized normal probability distribution is normally distributed with mean $\mu = 0$ and standard deviation $\sigma = 1$. In fact (cf. Equation 7.7), $p(0 + a \cdot 1 \leqslant z \leqslant 0 + b \cdot 1) = p(a \leqslant z \leqslant b) = A_n(a, b)$.

Appendix D gives the values of $p(0 \leqslant z \leqslant c) = N(c)$ for values of c. Example 9 shows how to compute probabilities $p(a \leqslant z \leqslant b)$ for all values of a and b.

9 Example

A variable z is normally distributed with mean 0 and standard deviation 1. Find the probabilities (a) $p(1.1 \leqslant z \leqslant 2.1)$, (b) $p(|z| \leqslant 2)$, (c) $p(z \geqslant 1.6)$, (d) $p(-1.2 \leqslant z \leqslant 2.4)$, and (e) $p(-1 \leqslant z \leqslant -.5)$.

Method. The value $p(1.1 \leqslant z \leqslant 2.1) = A_n(1.1, 2.1)$ is the area between $z = 1.1$ and $z = 2.1$ and may be found as the *difference* between $A(0, 2.1)$ and $A(0, 1.1)$:

$$A_n(1.1, 2.1) = N(2.1) - N(1.1)$$

From Appendix D we obtain

a. $p(1.1 \leqslant z \leqslant 2.1) = N(2.1) - N(1.1) = .4821 - .3643 = .1178$

To find $p(|z| \leqslant 2) = p(-2 \leqslant z \leqslant 2)$, we use symmetry to find $A_n(-2, 2)$. Using Appendix D we obtain

b. $p(|z| \leqslant 2) = A_n(-2, 2) = 2N(2) = (2)(.4772) = .9544$

To find $p(z \geqslant 1.6) = p(1.6 \leqslant z < \infty)$, we note that $A_n(0, \infty) = .5000$. Thus $A_n(1.6, \infty) = .5000 - N(1.6)$, and

c. $p(z \geqslant 1.6) = A_n(1.6, \infty) = .5000 - .4452 = .0548$

The value of $p(-1.2 \leqslant z \leqslant 2.4)$ is best found by finding the areas $A_n(-1.2, 0)$ and $A_n(0, 2.4)$ and adding. We find the former area by symmetry: $A_n(-1.2, 0) = A_n(0, 1.2) = N(1.2)$. Thus

d. $p(-1.2 \leqslant z \leqslant 2.4) = N(1.2) + N(2.4)$
$= .3849 + .4918 = .8767$

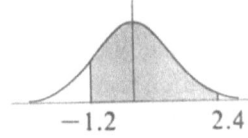

Finally, $A_n(-1, -.5) = A_n(.5, 1)$ by symmetry. Using the method in part a this value is $N(1) - N(.5)$. Thus

e. $p(-1 \leqslant z \leqslant -.5) = N(1) - N(.5) = .3413 - .1915$

$= .1498$

One of the important properties of a normally distributed variable is that the values lie close to the mean with a probability significantly higher than our previous conservative Chebyshev estimates.

10 Example

Suppose z is normally distributed, with mean 0 and standard deviation 0. For what value c is $p(|z| \leqslant c) = .50$. Similarly, find values c for which $p(|z| \leqslant c) = .90, .95$, and $.99$.

Method. By symmetry, $p(|z| \leqslant c) = p(-c \leqslant z \leqslant c) = 2N(c)$. Thus the problem amounts to solving the equation $2N(c) = .50, 2N(c) = .90$, etc.

For $2N(c) = .50$, $N(c) = .2500$, and we find from Appendix D that $c = .67$. For $2N(c) = .90$, $N(c) = .4500$, and $c = 1.65$. Similarly, $2N(c) = .95$ yields $c = 1.96$ and $2N(c) = .99$ yields $c = 2.58$. Summarizing,

$$p(|z| \leqslant \ .67) = .50$$
$$p(|z| \leqslant 1.65) = .90$$
$$p(|z| \leqslant 1.96) = .95$$
$$p(|z| \leqslant 2.58) = .99$$

Thus half the area under the standardized normal curve is in the range $|z| \leqslant .67$. Similarly, 95 percent of the area is in the range $|z| \leqslant 1.96$. For a normally distributed variable x with arbitrary mean μ and standard deviation σ, we can say that x is within $.67\,\sigma$ of μ with probability .50. Similarly, $p(|x - \mu| \leqslant 1.96\,\sigma) = .95$, etc.

It is worth comparing the values $p(|x - \mu| \leqslant c\sigma) = p(|z| \leqslant c)$ for a *normally distributed variable* with the bounds given by Chebyshev's theorem. Chebyshev's theorem states that $p(|x - \mu| \leqslant z\sigma)$ is at least $1 - (1/z^2)$. (We proved this for a discrete distribution attaining only a finite number of values $x = x_1, \ldots, x_n$.) We have pointed out that this is a conservative upper bound, because it is valid for all random variables. Let us now compare this estimate with actual values for a *normal* distribution.

We have $p(|z| \leqslant 2) = (2)\ (.4772) = .9544$. Chebyshev's theorem gives $1 - \frac{1}{4} = .7500$ as a lower bound. Similarly, $p(|z| \leqslant 3) = .9974$, with a Chebyshev bound of $1 - \frac{1}{9} = .89$. *Thus, if x is normally distributed $|x - \mu| \leqslant 3\sigma$ with probability 99.7 percent. The best that could be said of a variable x with any distribution is that $|x - \mu| \leqslant 3\sigma$ with probability 88.9 percent. The* values of $p(|z| \leq c)$ for a standardized normally distributed variable z (mean 0, standard deviation 1) are plotted for various values of c and are compared with the Chebyshev bounds in Fig. 7.10.

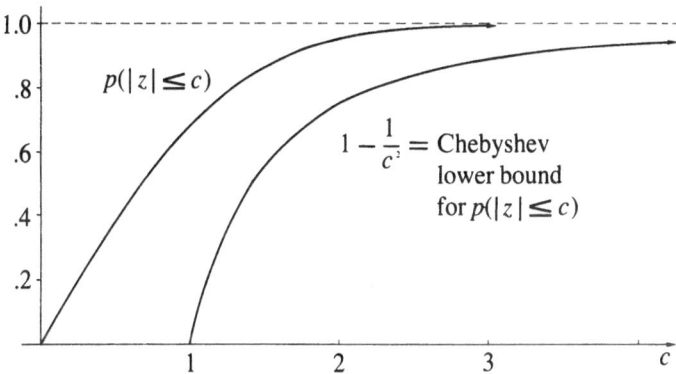

7.10 $p(|z| \leq c)$ for a Normal Distribution

11 Example

The measurement d of the diameter of a precision ball bearing is known to be normally distributed with $\mu = 3.40$ cm and $\sigma = .03$ cm. a. Find the probability that a measurement is larger than 3.35 cm. b. Find the probability that a measurement is between 3.33 and 3.43.

Method. Using Equation 7.8 (with d instead of x), we obtain

$$d = 3.40 + .03z$$

Then the condition $d > 3.35$ can be expressed as a condition on z:

$$d > 3.35$$
$$3.40 + .03z > 3.35$$
$$.03z > -.05$$
$$z > -5/3 = -1.67$$

Since z is normally distributed (mean 0, standard deviation 1), $p(d > 3.35) = p(z > -1.67) = .5000 + .4525 = .9525$. Similarly, the condition $3.33 \leq d \leq 3.43$ is expressible as an inequality in z:

$$3.33 \leq 3.40 + .03z \leq 3.43$$
$$-.07 \leq .03z \leq .03$$
$$-2.67 \leq z \leq 1$$

Thus

$$p(3.33 \leq d \leq 3.43) = p(-2.67 \leq z \leq 1) = .4962 + .3413 = .8375$$

Thus $d > 3.35$ with probability .9525, and d is between 3.33 and 3.43 with probability .8375.

12 Example

A factory fills sacks of fine sand. The amount of sand is known to average 51 lb with a standard deviation of $\frac{1}{2}$ lb. Assuming that the amount of sand in a sack is normally distributed, what percentage of the factory's output will weigh 50 lb or more?

Method. The amount s of sand is normally distributed with $\mu = 51$, $\sigma = \frac{1}{2}$. We wish to find $p(s \geqslant 50)$. Using Equation 7.8 with s instead of x, we set $s = 51 + \frac{1}{2}z$. Thus

$$p(s \geqslant 50) = p(51 + \tfrac{1}{2}z \geqslant 50)$$
$$= p(\tfrac{1}{2}z \geqslant -1)$$
$$= p(z \geqslant -2)$$

Since z is normally distributed, we compute

$$p(z \geqslant -2) = .5000 + .4772 = .9772 = 97.72\%$$

EXERCISES

1. Suppose the variable z is normally distributed with $\mu = 0, \sigma = 1$. Using Appendix D find

    a. $p(2 \leqslant z \leqslant 2.5)$ **b.** $p(z \geqslant -1.2)$
    c. $p(.91 \leqslant z \leqslant 1.11)$ **d.** $p(z^2 \leqslant 2)$

2. Suppose the variable x is normally distributed with mean 95 and standard deviation 4. Find

    a. $p(x \geqslant 94)$ **b.** $p(90 \leqslant x \leqslant 99)$
    c. $p(x \geqslant 105)$ **d.** $p(94 < x < 96)$

3. The variable x is normally distributed with mean 10 and standard deviation .2. Find a range of values for x so that x will fall in this range with probability .9. Similarly, find a range of values for x in which x will fall with probability .8.

4. The rocks in a pile have a mean diameter 15 mm and a standard deviation of 2 mm. The rocks are put through a sieve which permits any rock smaller than 17.5 mm in diameter to pass through but which stops rocks of 17.5 mm or larger. (Assume that the diameters are normally distributed.) What percentage of rocks pass through?

5. The grades on an examination are normally distributed with mean 70 percent and standard deviation 7 percent.

    a. One thousand students take the exam. On the average, how many students will receive a grade of 85 percent or higher?
    b. It is desired to determine a passing grade so that 90 percent of the students will pass. What should the passing grade be?

c. If 15 percent of the students should receive A's, what grade should receive an A?

6. On the average, 90 faculty members will attend a faculty meeting, with a standard deviation of 6. (Assume that the attendance is normally distributed.) What is the probability that 80 or more faculty members turn up at the next meeting?

7. For the faculty in Exercise 6 it is felt that no meeting should be held unless at least q members show up (q = quorum), lest democracy not prevail. (The value q is to be determined.) On the other hand, in the interest of efficiency, as well as boredom, it is considered always desirable to hold faculty meetings. Thus democracy and efficiency, respectively, act to increase and to decrease the correct value of q. A compromise is worked out whereby it is agreed that the faculty shall fail to have a quorum with probability $\frac{1}{10}$. What value of q should be used? If you want the faculty to meet rarely, say with probability $\frac{1}{10}$, what value of q should be used?

3 BINOMIAL DISTRIBUTION

The binomial distribution is one of the most commonly used discrete distributions. We shall describe it by an example involving coin tossing.

13 Example

A coin has probability p of landing heads and probability $q = 1 - p$ of landing tails. If n of these coins are tossed independently, find
   a. the distribution for the number of heads that turn up.
   b. the mean number of heads.
   c. the standard deviation for the number of heads.

Method. We let X = number of heads. [X is a random variable on the space S^n, where $S = \{H, T\}$ and $p(H) = p$, $p(T) = q$.] Clearly, the possible values of X are $x = 0, 1, 2, \ldots, n$. We denote the distribution of X by $b(x; n, p)$ to bring the given fixed numbers n and p explicitly into the notation. [Thus $b(x; n, p)$ = probability that $X = x$, or simply the probability of x heads.] We are required to find $b(x; n, p)$ for $x = 0, 1, \ldots, n$. But $b(0; n, p) =$ probability of no heads $= p(TT\ldots T) = q^n$, since $p(T) = q$, and the tosses are independent. To find $b(1; n, p)$, note that one head can occur in n distinct ways: $HTT\ldots T$, $THTT\ldots T, \ldots, TT\ldots TH$. Each of these ways has probability pq^{n-1}. Thus $b(1: n, p) = p(X = 1) =$ probability of one head $= npq^{n-1}$.

In general, suppose that we are given an integer x ($0 \leqslant x \leqslant n$) and we wish to find $p(X = x) = b(x; n, p)$. An outcome that has exactly x heads (and

$n-x$ tails) can be found by writing a sequence of x heads and $n-x$ tails. Such an outcome has probability $p^x q^{n-x}$ because we multiply probabilities (the tosses are independent) and each head yields a factor p and each tail a factor q. There are as many outcomes with x heads as there are ways to choose x places from among the n in which to locate the H's. Thus there are $\binom{n}{x}$ outcomes, and the probability of x heads is

$$b(x; n, p) = \binom{n}{x} p^x q^{n-x} \qquad (x = 0, 1, \ldots, n) \qquad (7.9)$$

We can illustrate for $n = 4$. The outcomes are

$x = 0$	$x = 1$	$x = 2$	$x = 3$	$x = 4$
TTTT	HTTT	HHTT	HHHT	HHHH
	THTT	HTHT	HHTH	
	TTHT	HTTH	HTHH	
	TTTH	THHT	THHH	
		THTH		
		TTHH		

The individual probabilities are

$$q^4 \qquad pq^3 \qquad p^2q^2 \qquad p^3q \qquad p^4$$

The number of outcomes are

$$\binom{4}{0} = 1 \quad \binom{4}{1} = 4 \quad \binom{4}{2} = 6 \quad \binom{4}{3} = 4 \quad \binom{4}{4} = 1$$

Thus the probabilities for $x = 0, 1, \ldots, 4$, are, respectively, $b(0; 4, p) = q^4$, $v(1; 4, p) = 4pq^3$, $b(2; 4, p) = 6p^2q^2$, $b(3; 4, p) = 4p^3q$, and $b(4;4, p) = p^4$.

To compute $E(X)$, the mean number of heads, and $\sigma(X)$, the standard deviation, it is convenient to use the technique of Sections 4 of Chapter 5 and 3 of Chapter 6. We let $N_1 =$ number of heads on the first toss, $N_2 =$ number of heads on the second toss, etc. Then the N_i are independent and each N_i has the same distribution, because

$$N_i = \begin{cases} 1 & \text{with probability } p \\ 0 & \text{with probability } q \end{cases}$$

Thus $E(N_i) = p \cdot 1 + q \cdot 0 = p$, while $\sigma^2(N_i) = E(N_i^2) - [E(N_i)]^2 = 1^2p + 0^2q - p^2 = p - p^2$. More simply, $\sigma^2(N_i) = p(1-p) = pq$. Then since

$$X = N_1 + \cdots + N_n$$

we have

$$E(X) = E(N_1) + \cdots + E(N_n)$$
$$\mu = E(X) = p + \cdots p$$
$$\mu = E(X) = np \tag{7.10}$$

Similarly, using Equation 6.28, we have

$$\sigma^2(X) = \sigma^2(N_1) + \cdots + \sigma^2(N_n)$$
$$= pq + \cdots + pq$$
$$= npq$$

Thus

$$\sigma = \sigma(X) = \sqrt{npq} \tag{7.11}$$

Equations 7.9, 7.10, and 7.11 constitute the complete answers to questions a, b, and c proposed above.

The values of $b(x; n, p)$ have been extensively tabulated. Appendix E is a brief table. The following examples illustrate its use.

14 Example

A seed has probability .9 of germinating when planted. Six seeds are planted. What is the probability that at least 5 of the seeds germinate?

Method. We regard the planting of a seed as the toss of a coin, and we regard "heads" as nongerminating with $p = .1$. Thus $q = .9$ is the probability of "tails," i.e., germination. [Appendix E gives values of $b(x; n, p)$ for $p = .1$, $.2, \ldots, .5$. Thus we arrange for "heads" to have probability $\leq .5$.] Under this interpretation of a planting as a coin toss, we have $n = 6$ "tosses" and $p = .1$. We want the probability of 5 or 6 "tails," so we compute the probability of 1 or 0 "heads." This is $b(0; 6, .1) + b(1; 6, .1)$. Going to Appendix E, under the column $p = .1$ and alongside $n = 6$ we find $b(0) = .5314$ and $b(1) = .3543$. The required probability is $.5314 + .3543 = .8857$.

Using Equation 7.9, we could explicitly write

$$b(0) + b(1) = (.9)^6 + (6)(.9)^5(.1)$$

However, this computation is somewhat long and the work has already been done in Appendix E.

The choice of .1 for p, rather than the more natural .9, can be avoided by using the formula

$$b(x; n, p) = b(n - x; n, 1 - p) \tag{7.12}$$

In fact, the left-hand side is the probability of x heads in a toss of n coins where $p(H) = p$. The right-hand side may be interpreted as the probability of $n - x$ tails in a toss of n coins where $p(T) = 1 - p$. Clearly these values are the same. More formally, we can prove Equation 7.12 directly using

Equation 7.9.

$$b(x; n, p) = \binom{n}{x} p^x (1-p)^{n-x}$$

and

$$b(n-x; n, 1-p) = \binom{n}{n-x}(1-p)^{n-x}[1-(1-p)]^{n-(n-x)}$$

$$= \binom{n}{x}(1-p)^{n-x} p^x = b(n; n, p)$$

proving Equation 7.12.

In the above example, had we used $p = .9 =$ probability of germination, we would find, with the help of Equation 7.12,

$$b(5; 6, .9) = b(6-5; 6, 1-.9) = b(1; 6, .1)$$

and

$$b(6; 6, .9) = b(6-6; 6, 1-.9) = b(0, 6, .1)$$

We computed $b(5; 6, .9) + b(6; 6, 19)$ by evaluating $b(1; 6, .1) + b(0; 6, .1)$.

The above example shows how a process may be interpreted as a coin toss. Some other examples are: The birth of a child may be regarded as a toss of a coin, with "boy" as "heads." If p is the probability of a boy being born, then $b(x; n, p)$ is the probability that, of n newly born babies in a hospital, exactly x are boys. The underlying assumptions, as in our coin tossing, are that (a) each birth has the same probability p of being a boy, and (b) the births are independent. Similarly, if 8 dice are tossed, the probability that 2 or less 5's are tossed is $b(2; 8, \frac{1}{6}) + b(1; 8, \frac{1}{6}) + b(0; 8, \frac{1}{6})$. Here the toss of a 5 is regarded as "heads," and $p = \frac{1}{6}$, $n = 8$.

In general, we say that we have *Bernoulli trials*, if a sequence of experiments is performed such that (a) the probability of success for each trial is the same number p, with the probability of failure equal to $q = 1-p$, and (b) the trials are independent. We can then simply state: *For n Bernoulli trials, the probability of exactly x successes* $(x = 0, 1, \ldots, n)$ *is* $b(x; n, p) = \binom{n}{x} p^x q^{n-x}$.

This formula simplifies a bit for $p = \frac{1}{2}$. In this case $q = 1-p = \frac{1}{2}$, and $p^x q^{n-x} = (1/2^x)(1/2^{n-x}) = 1/2^n$. Thus

$$b(x; n, \tfrac{1}{2}) = \binom{n}{x}/2^n \tag{7.13}$$

(Compare Example 25 of Chapter 2.)

We now put these considerations in the form of a definition.

15 Definition

Let $X =$ number of successes in n Bernoulli trials, where $p =$ probability of a success on any individual trial. The distribution of X, denoted $b(x; n, p)$

or simply $b(x)$ if n and p are understood, is called the *binomial distribution* (with parameters n and p). The term $b(x)$ is given by the formula

$$b(x; n, p) = \binom{n}{x} p^x q^{n-x} \qquad (x = 0, 1, \ldots, n) \qquad (7.9)$$

As in Example 12, the mean μ and standard deviation σ for X is given by

$$\mu = np \qquad (7.10)$$
$$\sigma = \sqrt{npq} \qquad (7.11)$$

The term "binomial" is used here because the expressions $\binom{n}{x} p^x q^{n-x}$ are precisely the terms in the binomial expansion of $(q+p)^n$:

$$(q+p)^n = q^n + nq^{n-1}p + \binom{n}{2}q^{n-2}p^2 + \cdots + \binom{n}{x}q^{n-x}p^x + \cdots + p^n$$

or

$$(q+p)^n = \sum_{x=0}^{n} b(x; n, p)$$

Note that since $q+p = 1$, we have

$$1 = \sum_{x=0}^{n} b(x; n, p) \qquad (7.14)$$

This is a special case of Equation 5.8,

$$1 = \sum p(x_i)$$

which is valid for all distributions.

16 Example
A multiple-choice test has 10 problems, each with 5 choices. Assuming pure guesswork on the part of a student taking the exam: a. Find the probability that the student gets a grade of 30 percent or more. b. Find the probability that he gets a grade of 50 percent or more. c. Find the average grade and standard deviation of the grade for a test done by pure guesswork.

Method. Regard "success" as guessing correctly. Then $p = \frac{1}{5} = .2$ and $n = 10$.

a. Writing $b(x)$ instead of $b(x; 10, .2)$, we must find that $b(3) + b(4) + \cdots + b(10)$. This is simply $1 - [b(0) + b(1) + b(2)]$, by Equation 7.14. Using Appendix E, under $p = .2$, $n = 10$, we find $b(0) + b(1) + b(2) = .1074 + .2684 + .3020 = .6778$. Thus the required probability is $1 - .6778 = .3222$.

b. The probability of a grade of 50 percent or more is $b(5) + b(6) + \cdots + b(10) = .0264 + .0055 + .0008 + .0001 = .0328$. The fact that this figure is so low will not be surprising to any student who guesses a lot on multiple-choice tests.

c. The mean number of correct guesses is $\mu = np = 10(.2) = 2$. The average grade is thus 20 percent.[3] Similarly, $\sigma = \sqrt{npq} = \sqrt{(10)(.2)(.8)} = \sqrt{1.6} = 1.26$. The standard deviation for the grade is 12.6 percent.

17   Example

It is claimed that 60 percent of the population of a large city favors condidate A. It is decided by his opponent, candidate B, to take a poll of 10 people at random to check this figure. Candidate B has decided to reject the 60 percent figure as ridiculous if his poll shows 4 or fewer people favoring candidate A. Is B a bit rash?

Method.   Let us assume that $p = .6$ is the probability of favoring A. Then we have Bernoulli trials with $p = .6$, $n = 10$.[4] Candidate B apparently considers the probability that $X \leq 4$ ridiculously low, so low that the occurrence of $X \leq 4$ is enough to convince him that the assumption $p = .6$ is wrong. Let us compute $p(X \leq 4)$. The probability of 4 or fewer people favoring A is $b(4; 10, .6) + b(3; 10, .6) + \cdots + b(0; 10, .6)$. By Equation 7.12 this is $b(6; 10, .4) + b(7; 10, .4) + \cdots + b(10; 10, .4)$, which is equal to $.1115 + .0425 + \cdots + .0001 = .1663$. This is not particularly small. (It is about 1 in 6.) The evidence would not generally be considered strong enough to reject the assumption $p = .6$.

Generally, statisticians and such take a small figure, say .05 or .01 if they are feeling *very* conservative, and use it as a measure of the probability of an extremely rare event. If they want to test a hypothesis (e.g., the 60 percent figure above), they will then devise a test that can succeed only with this small probability, given that hypothesis. If the test succeeds, they will reject the hypothesis. In the above example candidate B was very rash, because a probability of .166 is not considered so small.

If the 5 percent rejection level were to be used, the above figures indicate that B should reject the hypothesis of 60 percent for A, if 2 or fewer people out of 10 favor A. (Three or fewer will work if he shades 5 to 6 percent.)

Another problem B might want solved is to estimate, with some precision, what percentage of the population actually does favor A. Here your intuition tells you that a random sample of 10 is too small to give the figure with any accuracy. But beware! Your intuition may also tell you that the samples used by national polling organizations and television polls are also too

---

3 Why? Can you justify this step from 2 to 20 percent?
4 Strictly speaking this is not *quite* correct. If one person is chosen, at random, and a different person is then chosen, the events will not be independent. If the population is 100,000 voters, if 60,000 favor A, and if the first person polled happens to favor A, then the probability for the next is, technically, 59,999/99,999. This is very close to, but not exactly, .6, and it will remain close for 10 trials. Thus we may regard this process as very accurately approximated by Bernoulli trials.

small. ("What," you will say, "only 1,000 people out of 1,000,000 and they want *me* to believe their prediction!") Hold off making such intuitive guesses until you finish reading this chapter.

EXERCISES

In the following exercises, use Appendix E wherever applicable.

1. Using Equation 7.9 or 7.13, check the accuracy of Appendix E by actually computing
    a. $b(3; 6, .5)$ b. $b(2; 4, .2)$ c. $b(3; 4, .1)$
    d. $b(3; 5, .3)$ e. $b(6; 10, .15)$.

2. A fair coin is tossed 10 times. What is the probability of 6 heads and 4 tails? What is the expected number of heads and the standard deviation for the number of heads?

3. A stamp dealer is known to send out defective stamps with probability .1. If 10 stamps are ordered, what is the probability that at most 1 defective stamp is delivered?

4. Assume that 30 percent of the adult population of a certain city are college graduates. Eight of these adults find themselves (independently and at random) in a doctor's office. Using Appendix E, find the probability that 3 or fewer are college graduates.

5. A college has accepted 700 students for admission. It knows by past experience that a student will actually decide to come to the college with probability 30 percent. What is the expected number of students who will come to the college, and with what standard deviation?

6. Assume that the chance of a twin birth occurring is $\frac{1}{90}$. If 200 births take place in a hospital during 2 weeks, what is the probability that 3 or more twin births occur? [Express your answer in terms of $b(x; p, q)$.] Do not evaluate. Give as simple an answer as possible.

7. A manufacturer of flashlights knows that his manufacturing process produces defective flashlights with probability .1. He sends out a shipment of 10 flashlights. What is the probability that 8 or more are not defective?

8. Hamsters infected with a certain disease are known to recover with probability 20 percent. A drug company claims that it has an effective cure for this disease, but you, as a good scientist, are skeptical. You decide to test the drug on 9 hamsters. How many hamsters will have to recover for you to admit they have a case? (*Note*: Be liberal, and consider that if an event with probability .10 or less occurs, you will lose your skepticism. Similarly, test the probabilities .05 and .01 as your measure of a rare event.)

9. Prove: For fixed n and p, $b(x+1) > b(x)$ for $x < \mu - q$ and $b(x+1) < b(x)$ for $x > \mu - q$. Thus we can roughly say that $b(x)$ increases with x until $x = \mu - q$, after which $b(x)$ decreases with x. [*Hint:* Consider the equation $b(x+1) > b(x)$, or $b(x+1)/b(x) > 1$.]

4 NORMAL APPROXIMATION

A glance at the graphs of various binomial distributions for large n indicates that the shape of each graph is remarkably similar to a normal curve. Figure 7.11 plots $b(x; 10, .3)$ for $x = 0, \ldots, 10$ in the form of a bar graph. Figure 7.12 plots $b(x; 12, .6)$ similarly. In each case the similarity to a normal curve is striking. It is one of the remarkable results in mathematics that, for large n,

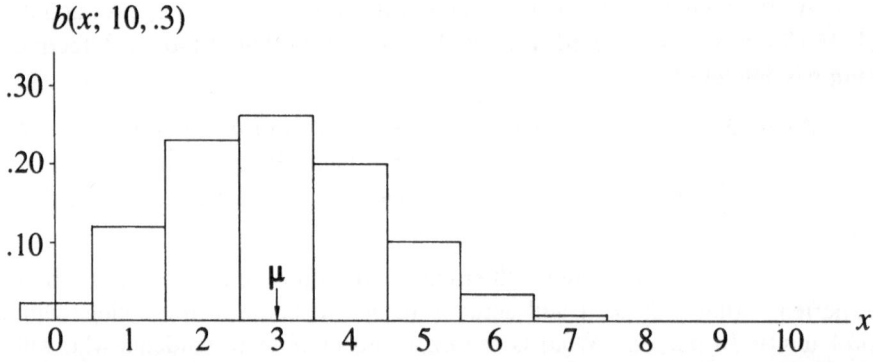

7.11 Graph of b(x; 10, .3)

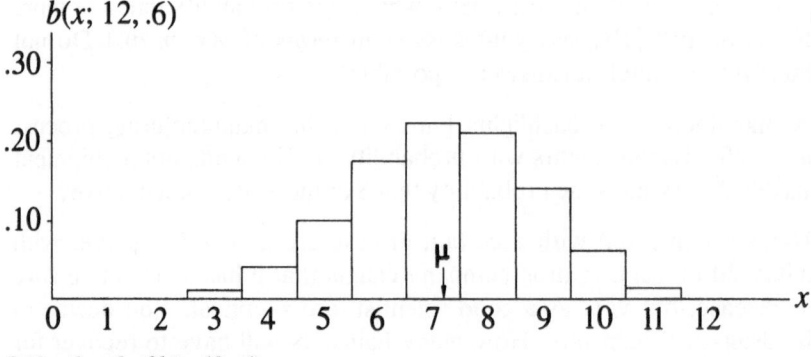

7.12 Graph of b(x; 12, .6)

the *binomial distribution* $b(x; n, p)$ can be closely approximated by a *normal* distribution with the same mean μ and standard deviation σ. We shall first state this theorem and then show how to use it. A proof is beyond the scope of this text.

18 Theorem

For large values of n, a binomial distribution can be approximated by a normal distribution with the same mean and standard deviation as the binomial distribution. More explicitly, suppose n and p are given, determining a binomial distribution with $\mu = np$ and $\sigma = \sqrt{npq}$. Then, if x_1 and x_2 are integers and if n is large, we have

$$\sum_{x=x_1}^{x_2} b(x; n, p) \approx p(x_1 - \tfrac{1}{2} \leqslant x \leqslant x_2 + \tfrac{1}{2}) \tag{7.15}$$

Here we are using $p(a \leqslant x \leqslant b)$ to denote the probability that the *real* variable x is between a and b, assuming that x is normally distributed with mean μ and standard deviation σ.[5]

The left-hand side of the approximation 7.15 is the probability that the *integer* variable x is between x_1 and x_2, if x is distributed according to the binomial distribution. The significance of the values $x_1 - \tfrac{1}{2}$ and $x_2 + \tfrac{1}{2}$, rather than x_1 and x_2 in the right-hand side of Equation 7.15 is explained on page 241. The approximation is illustrated in Fig. 7.13, in which the graph of $b(x; 16, .5)$ is given along with the graph of the probability density of a normally distributed variable with the same mean $\mu = 8$ and standard deviation $\sigma = 2$. The probabilities $\sum_{x=4}^{10} b(x)$ and $p(3\tfrac{1}{2} \leqslant x \leqslant 10\tfrac{1}{2})$ are both illustrated as areas.

In practice, the approximation formula 7.15 works even when n is not too large, say as small as 10. The precise statement of Theorem 18 is given by a statement involving limits. One form of this is

$$\sum_{\mu+z_1\sigma \leqslant x \leqslant \mu+z_2\sigma} b(x; n, p) \to A(z_1, z_2) \qquad \text{as } n \to \infty$$

where $A(z_1, z_2)$ is the area under the normal curve $y = n(z)$ between $z = z_1$ and $z = z_2$.

19 Example

Forty percent of the population of a city is opposed to city parades. Ten people are sampled. What is the probability that between 3 and 5 of the people polled (inclusive) oppose parades? Approximate the answer using the normal distribution.

---

5 For the remainder of the text we shall use the notation $p(a \leqslant x \leqslant b)$ only for a normally distributed variable x. As usual, z will denote a normally distributed variable with mean 0 and standard deviation 1.

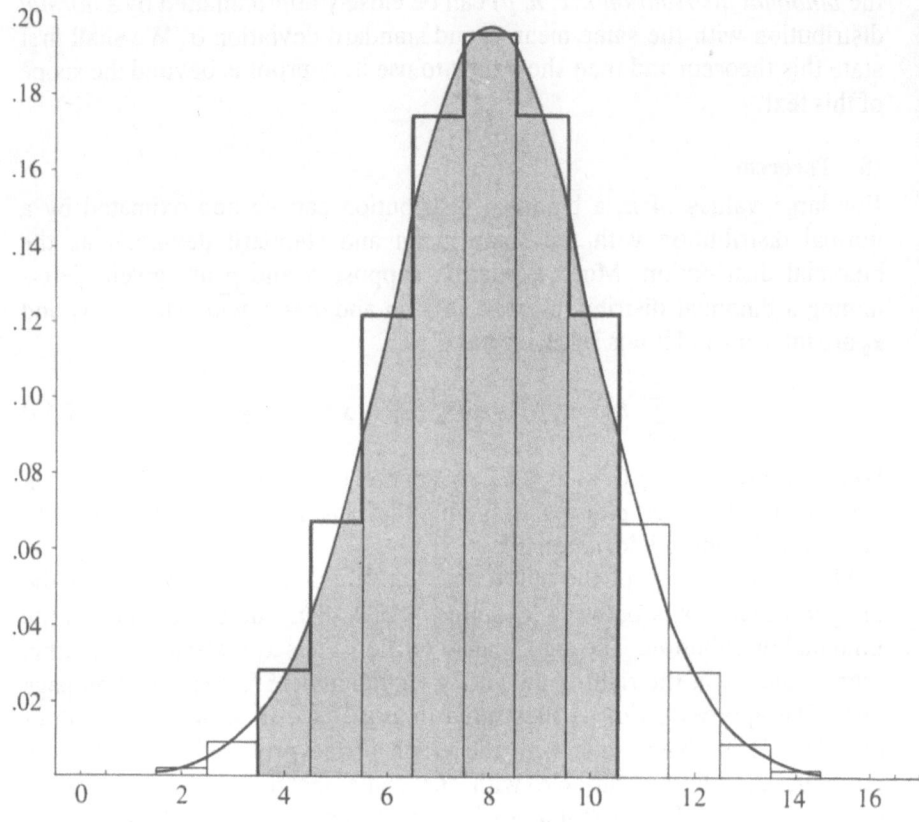

7.13   Comparison of Binomial and Normal Distribution

Method.   We can assume a binomial distribution with $n = 10$, $p = .4$. (See the footnote on page 254.) The required answer is clearly $b(3) + b(4) + b(5)$, where we are writing $b(x) = b(x; 10, .4)$. Using Appendix E, the answer is $.2150 + .2508 + .2007 = .6665$.

In this example the normal approximation procedure is far longer than this direct method. However, it illustrates the procedure well. We have $n = 10$, $p = .4$. Thus $\mu = np = 4$, and $\sigma = \sqrt{(10)(.4)(.6)} = \sqrt{2.4} = 1.55$ (to 2 decimal places). The required probability that $3 \leqslant x \leqslant 5$ (x an integer) is replaced by the probability $p(2.5 \leqslant x \leqslant 5.5)$, where x is real and normally distributed with $\mu = 4$, $\sigma = 1.55$. Setting $x = 4 + 1.55z$, the inequality

$$2.5 \leqslant x \leqslant 5.5$$

becomes

$$2.5 \leqslant 4 + 1.55z \leqslant 5.5$$
$$-1.5 \leqslant 1.55z \leqslant 1.5$$
$$-.97 \leqslant z \leqslant .97$$

The required probability, using Appendix D, is

$$2N(.97) = (2)(.3340) = .6680$$

a difference of only .0015 from the exact answer, .6665.

The accuracy of the normal approximation has thus been illustrated for the rather moderately sized $n = 10$.

20 Example

Two hundred fair coins are independently tossed. What is the probability that the number of heads is between 90 and 115 inclusive?

Method. We have a binomial distribution with $n = 200$, $p = \frac{1}{2}$. Thus $\mu = np = 100$, and $\sigma = \sqrt{npq} = \sqrt{50} = 7.07$. The required probability that $90 \leqslant x \leqslant 115$ is approximated by $p(89.5 \leqslant x \leqslant 115.5)$, where x is normally distributed with $\mu = 100$ and $\sigma = 7.07$. Setting $x = 100 + 7.07z$, we have the following equivalent inequalities:

$$89.5 \leqslant x \leqslant 115.5$$
$$89.5 \leqslant 100 + 7.07z \leqslant 115.5$$
$$-10.5 \leqslant 7.07z \leqslant 15.5$$
$$-1.49 \leqslant z \leqslant 2.19$$

But

$$p(-1.49 \leqslant z \leqslant 2.19) = N(2.19) + N(1.49) = .4857 + .4319 = .9176$$

Thus the required probability is approximately .9176.

21 Example

A fair die is tossed 100 times. What is the probability that the number of 6's which turn up is 15 or more?

Method. The number x of 6's is governed by a binomial distribution with $n = 100$, $p = \frac{1}{6}$. Thus $\mu = np = 16.67$ and $\sigma = \sqrt{(100)(\frac{1}{6})(\frac{5}{6})} = (\frac{10}{6})\sqrt{5} = 3.73$. The probability that x is 15 or more is $b(15) + b(16) + \cdots$. The normal approximation is simply $p(14.5 \leqslant x < \infty)$.[6] Setting $x = 16.67 + 3.73z$, the inequality

$$14.5 \leqslant x$$

becomes

$$14.5 \leqslant 16.67 + 3.73z$$
$$-2.17 \leqslant 3.73z$$
$$-.58 \leqslant z$$

Using Appendix D, $p(z \geqslant -.58) = .7190$. Thus the required probability is approximately .7190.

---

6 Note that 100.5 is the proper upper limit. This is over 22σ larger than μ and can therefore be replaced by ∞.

22   Example

A fair coin is tossed 100 times. Find a number N such that the probability of N or more heads is approximately .90.

Method.   Here $n = 100$, $p = \frac{1}{2}$. Thus $\mu = 50$, $\sigma = \sqrt{(100)(\frac{1}{2})(\frac{1}{2})} = 5$. The probability of N or more heads is $b(N) + b(N+1) + \cdots \approx p(N - \frac{1}{2} \leqslant x)$. Set $x = 50 + 5z$. We have the following equivalent inequalities.

$$N - \tfrac{1}{2} \leqslant x$$
$$N - \tfrac{1}{2} \leqslant 50 + 5z$$
$$N - 50.5 \leqslant 5z$$
$$\frac{N - 50.5}{5} \leqslant z$$

Thus

$$p(N - \tfrac{1}{2} \leqslant x) = p\left(\frac{N - 50.5}{5} \leqslant z\right) = A_n\left(\frac{N - 50.5}{5}, \infty\right)$$

We wish this latter probability to be .90. We therefore try to find a value c such that

$$A(c, \infty) = .90$$

From Appendix D, $c = -1.28$. Thus we set

$$\frac{N - 50.5}{5} = -1.28$$
$$N - 50.5 = -6.40$$
$$N = 44.1$$

Since N must be an integer, we choose $N = 44$. The probability that $x \geqslant 44$ will be slightly larger than .90.

A somewhat different way of proceeding is as follows: From Appendix D we find that

$$A(-1.28, \infty) = .90$$

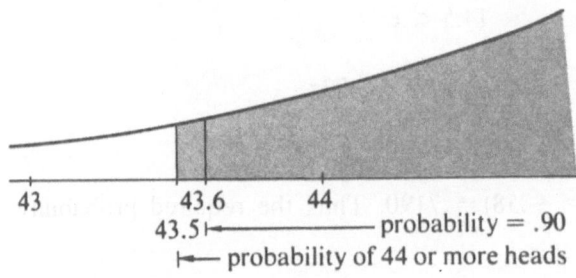

probability $= .90$

probability of 44 or more heads

7.14   Choosing N

Thus 90 percent of the area under the normal curve is to the right of -1.28. Translating this fact to a normal distribution with $\mu = 50$ and $\sigma = 5$, 90 percent of the area under this curve is to the right of $\mu - 1.28\sigma = 50 - (5)$ $(1.28) = 43.6$. But, referring to Fig. 7.14, the probability that $x \geqslant 44$ is approximately the area to the right of 43.5, which is somewhat larger than .90. Thus we choose $N = 44$.

EXERCISES

Where appropriate, use the normal approximation to the binomial distribution.

1. A coin is tossed 100 times. What is the probability that the relative frequency of heads is between 45 and 55 percent inclusive? Find this probability if the coin is tossed 200 times. Similarly, find the probability if the coin is tossed 500 times.

2. In Example 1 of Chapter 1, 3 dice were tossed 100 times. It was observed, experimentally, that the high number on the dice was 5 in 33 percent of the experiments. In Example 6 of Chapter 2 we found that the probability of a high of 5 was 28.2 percent. Find the probability that in a run of 100 experiments, the high of 5 will not occur between 24 and 32 times, inclusive.

3. A die is tossed 120 times. What is the probability that a 5 turns up between 15 and 25 times, inclusive? For what range of values can you predict that a 5 will turn up with probability .90 or more? .95 or more? .99 or more?

*4. A coin is known to be fair. It is desired to toss this coin N times. Choose N so that the relative frequency of heads will be between .49 and .51 inclusive, with probability .95 or more. Similarly, find N if the relative frequency of heads is to be between .48 and .52 with probability .98 or more.

5. A plant seed is known to germinate with probability .90. If 100 seeds are planted, what is the probability that 87 or more seeds germinate?

6. In Exercise 5 it is essential to have 100 or more germinated seeds. How many seeds should be planted to guarantee this with a 99 percent probability?

7. Assume that 6,000 people in a town favor candidate A and 4,000 people favor candidate B. A poll is taken of 20 people. (The people are chosen independently and at random.) What is the probability that 10 or more people polled favor candidate B? What is the probability that 15 or fewer people favor A?

8. Each student in a class of 30 is told to toss 5 coins and report whether or not 4 heads and 1 tail occurred. Each student performs this experiment 3 times, for a total of 90 experiments. What is the probability that between 12 and 16 occurrences, inclusive, of a 4-head and 1-tail split are reported.

9. A person claims that he can guess the suit of a card chosen at random from a deck. You test him 50 times, each time carefully shuffling the deck. Assuming that his claim is false, and that the person guesses correctly with probability $\frac{1}{4}$, what is the probability that he guesses correctly 18 or more times? Find an N so that the probability of N or more correct guess is about .10.

10. A multiple-choice test has 100 questions on it, each with 5 choices. Assuming random answers, what is the probability of achieving a grade of 25 percent or higher? Set a grade that will automatically fail 98 percent of the guessers.

11. In Exercise 10 suppose that all the questions have two answers which are obvious nonsense and which are eliminated before guessing randomly on the other 3. What is the probability of a grade of 40 percent or higher? Of 50 percent or higher? If someone gets a grade of 53 percent on this exam, present an argument proving that the person knows something about the subject.

12. Do Exercise 8 of Section 6.3 using the normal approximation to the binomial distribution.

5 STATISTICAL APPLICATIONS

One of the current myths about statisticians is that "statistics can prove anything." Another opposing myth is that "statistics can prove nothing." We now consider some statistical applications of the results of this chapter to learn more about the applications of probability theory. In all cases the idea behind these methods is that if our assumptions show that an event has a very small probability but nevertheless occurs in an experiment, then we may have cause to reject our assumptions.

23 Example
You are going to test a die to see if it is honest. You decide to test whether a 6 turns up about $\frac{1}{6}$ of the time, as it should if the die is honest. You toss the die 600 times, independently, and you observe that a 6 turned up 120 times.

You are undecided. Is 120 too large, or is 120 reasonably close to the expected number (100) of 6's?

Method. We have no business performing an experiment and waiting for the results before deciding what to do with them! However, this is a typically human error, so let us see what we might have done.

Prior to the experiment you decide to toss the die 600 times. *Assuming $p = \frac{1}{6}$ (the probability of a 6), you have a binomial distribution with $\mu = np = 100$ and $\sigma = \sqrt{npq} = \sqrt{(600)(\frac{1}{6})(\frac{5}{6})} = \sqrt{83.3} = 9.13$.* Therefore, you expect the number x of successes to be "close" to $\mu = 100$. If x is too far from 100, you will reject the assumption that $p = \frac{1}{6}$.

We must now decide how far x should be from 100 in order to reject our assumption that $p = \frac{1}{6}$. Let us find an interval about 100 large enough so that x will fall in this interval with probability 95 percent $= .9500$. (The figure 95 percent is an arbitrary decision on our part.) According to Appendix D, $N(1.96) = .4750 = 47\frac{1}{2}$ percent. Thus $A_n(-1.96, 1.96) = .9500$, and 95 percent of the area under the normal curve falls between $z = -1.96$ and $z = 1.96$. In terms of an arbitrary normal curve, we have $p(\mu - 1.96\sigma \leqslant x \leqslant \mu + 1.96\sigma) = .9500$. In our case $\mu = 100$, $\sigma = 9.13$. Since $1.96 \times 9.13 = 17.9$, we have $p(82.1 \leqslant x \leqslant 117.9) = .9500$. The probability that the number x of 6's is between 82 and 118 is approximately $p(81.5 \leqslant x \leqslant 118.5)$, hence greater than .95. We have every right to expect x to be between 82 and 118, because the probability that this occurs is slightly larger than .95. Thus $x < 82$ or $x > 118$ with very small probability (smaller than .05). Therefore, our test is as follows: If x is not within the interval $82 \leqslant x \leqslant 118$, we shall reject the hypothesis that $p = \frac{1}{6}$. In that case we say that we *reject the hypothesis $p = \frac{1}{6}$ at the 5 percent significance level.*

In particular, because 6 occurred 120 times in our test, we will say that the die is unfair. In doing so we admit that the die may be fair and that an event of small probability may have happened. If we want to be very conservative in our rejection, we may decide to reject at the 2 percent, or even 1 percent, level. In this case the number 1.96, which satisfied $p(|z| \leqslant 1.96) = .95$, is replaced by 2.33 $[p(|z| \leqslant 2.33) = .98]$ or by 2.58 $[p(|z| \leqslant 2.58) = .99]$, and we shall reject the hypothesis only when x is outside appropriately wider limits.

24 Example

Thirty hamsters have a disease from which they ordinarily recover with probability .4. A company claims that their remedy will increase the recovery rate to .8. Devise a test to decide on the company's claim.

Method. Suppose we feed the hamsters the medicine and see what happens. But first, we shall make up a test, considering only the two alternatives

$p = .4$ (that the remedy has no value at all) and $p = .8$ (as claimed by the manufacturer).

1. If $p = .4$, $n = 30$, then $\mu = 12$, $\sigma = \sqrt{npq} = \sqrt{7.2} = 2.68$. Thus, if x is the number that recover, and if x is "much" larger than 12, we shall reject the hypothesis that $p = .4$. (*Note*: We are deciding to consider whether x is much larger than 12 rather than whether x is very far away from 12, above or below.) Let us choose the 5 percent significance level. We find c with $N(c) = .4500$. In that case $p(z \leqslant c) = .9500$ and $p(z > c) = .05$. Using Appendix D we find $c = 1.65$. Thus, for a normally distributed x, $p(x > \mu + 1.65\sigma) = p(z > 1.65) = .05$. For $\mu = 12$, $\sigma = 2.68$, we have $\mu + 1.65\sigma = 12 + 4.43 = 16.63$, and therefore $p(x > 16.63) = .05$. If x is the number of recoveries, the probability that $x \geqslant 18$ is approximately $p(x \geqslant 17.5)$, hence smaller than .05. Thus, *at the 5 percent significance level, we reject* $p = .4$ if it turns out that $x \geqslant 18$.

2. If $p = .8$, as claimed, $n = 30$, then $\mu = 24$ and $\sigma = \sqrt{4.8} = 2.30$. If x hamsters recover, where x is "much" less than 24, we shall reject the hypothesis $p = .8$. Choosing the 5 percent significance level, we find as above that $p(z < -1.65) = .05$. (Thus 95 percent of the time z will be greater or equal to -1.65.) Since $\mu = 24$, $\sigma = 2.30$, the value $z = -1.65$ corresponds to $x = 24 - (2.30)(1.65) = 20.20$. Thus $p(z \leqslant 20.20) = .05$. If x is the number of recoveries, the probability that $x \leqslant 19$ is approximately $p(x \leqslant 19.5)$, hence smaller than .05. Thus *at the 5 percent significance level, we reject* the hypothesis $p = .8$ if it turns out that $x \leqslant 19$.

We summarize our test as follows. We feed the remedy to the hamsters and count the number x of recoveries. If $x \geqslant 18$, we affirm that $p \neq .4$. (The medicine has some effect.) If $x \leqslant 19$, we shall assert that the claim $p = .8$ is false. In each case we admit that even if $p = .4$ or if $p = .8$ there will be a 5 percent probability that we are wrong. See Fig. 7.15 for a graphical description of this test.

What happens if, say, only 6 hamsters survive? We then suspect that the

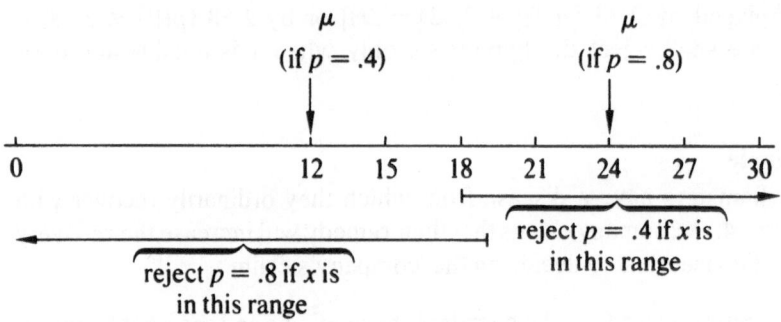

7.15   *A Test*

medicine has a negative effect. We go back to the drawing boards and devise a new test! Our procedure must always be determined first to prevent after-the-fact wishful thinking that might wrongfully color our thinking.

Our last examples show how to estimate population. We first show that, for a large sample, the observed relative frequency F of success is approximately normally distributed.

25 Theorem

Let n Bernoulli trials be performed, where the probability of success on each trial is p. Let F be the relative frequency of success. (F is a random variable.) Then

$$\mu_F = E(F) = p \tag{7.16}$$

$$\sigma_F = \sigma(F) = \sqrt{pq/n} \tag{7.17}$$

Furthermore, if n is large, the distribution of F may be approximated by a normal distribution with mean μ_F and standard deviation σ_F. Thus $a \le F \le b$ with probability approximately $p(a \le f \le b)$, where f is normally distributed with mean p and standard deviation $\sqrt{pq/n}$.

Proof. If X is the number of successes, we have

$$F = \frac{1}{n} X$$

Thus, using Equations 7.10 and 7.11 for the binomial distribution,

$$\mu_F = E(F) = E\left(\frac{1}{n}X\right) = \frac{1}{n}E(X) = \frac{1}{n} \cdot np = p$$

$$\sigma_F = \sigma(F) = \sigma\left(\frac{1}{n}X\right) = \frac{1}{n}\sigma(X) = \frac{1}{n}\sqrt{npq} = \sqrt{pq/n}$$

Finally, let us find the probability that $a \le F \le b$. Since $F = (1/n)X$, this inequality is equivalent to $a \le (1/n)X \le b$ or $na \le X \le nb$. If we set $x = np + z\sqrt{npq}$, this latter inequality has probability approximately

$$p(na - \tfrac{1}{2} \le np + z\sqrt{npq} \le nb + \tfrac{1}{2}) = p\left(\frac{na - np - \tfrac{1}{2}}{\sqrt{npq}} \le z \le \frac{nb - np + \tfrac{1}{2}}{\sqrt{npq}}\right)$$

where z has a standard normal distribution. Since n is large, we can ignore the small summands $\pm (\tfrac{1}{2}/\sqrt{npq})$, and the probability is approximately

$$p\left(\frac{na - np}{\sqrt{npq}} \le z \le \frac{nb - np}{\sqrt{npq}}\right) = p(na \le np + \sqrt{npq}\,z \le nb) =$$

$$p(a \le p + \sqrt{pq/n}\,z \le b)$$

Thus the probability that $a \le F \le b$ is approximately $p(a \le p + \sqrt{pq/n}\,z \le b)$, which is precisely the probability $p(a \le f \le b)$, where f is a normally distributed variable with mean p and standard deviation $\sqrt{pq/n}$. This completes the proof.

26 Example

An experiment has probability .3 of success. Approximately how many times should the experiment be performed in order that the relative frequancy F of success be between .28 and .32 with probability .9?

Method. F has mean $\mu = .30$. The condition $.28 \leqslant F \leqslant .32$ may be written $|F - .30| \leqslant .02$. According to Appendix D, $|z| \leqslant 1.65$ with probability .9. Thus, for any normal variable f, $|f - \mu| \leqslant 1.65\sigma$ with probability .9. We therefore arrange for $1.65\sigma = .02$, and we will have $|f - .30| \leqslant .02$ with probability .9. But $\sigma = \sqrt{pq/n} = \sqrt{(.3)(.7)/n}$. Thus we choose n so that

$$1.65\sigma = .02$$

$$1.65\sqrt{.21/n} = .02$$

$$\frac{(1.65)^2(.21)}{n} = .02^2$$

$$n = \frac{(1.65)^2(.21)}{.02^2} = 1{,}429.3$$

Thus we may choose $n \approx 1{,}430$.

Before going to our last example, we give a simple result that will save us some work later.

27 Theorem

For a binomial distribution with parameters n and p, the largest standard deviation possible is $\frac{1}{2}\sqrt{n}$, and it occurs when $p = \frac{1}{2}$. Similarly, the relative frequency F of success has standard deviation at most $\frac{1}{2}\sqrt{1/n}$.

Proof. The standard deviation for the binomial distribution is $\sqrt{npq} = \sqrt{np(1-p)}$. Therefore, we wish to find the largest value of $y = p(1-p)$ $(0 \leqslant p \leqslant 1)$. The graph of $y = p - p^2$ is given in Fig. 7.16. It is seen there that the largest value occurs when $p = \frac{1}{2}$, for which $p(1-p) = \frac{1}{4}$. Thus $\sigma = \sqrt{np(1-p)}$ is always less than or equal to $\sqrt{n \cdot \frac{1}{4}} = \frac{1}{2}\sqrt{n}$. Similarly, $\sqrt{pq/n}$ is at most $\frac{1}{2}\sqrt{1/n}$. This is the result. A more algebraic proof is outlined in the Exercises.

28 Example

You are polling a city of 1,000,000 voting adults to determine whether they favor candidate A or candidate B. You poll 500 such people (independently and at random) and you find that 280 favor A and 220 favor B. What announcement can you give to the press?

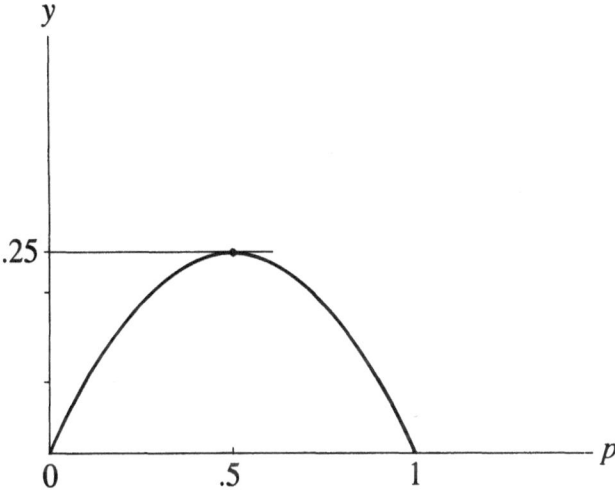

7.16   Graph of y = p − p²

Method.   The methodology is wrong. *First,* you plan the experiment and *then* you decide what to do with your results. Here is a plan.

If p is the fraction of adults favoring A and q is the fraction favoring B, our poll may be thought of as $n = 500$ Bernoulli trials. The relative frequency F of those favoring A has mean p and standard deviation $\sqrt{pq/500}$. According to Theorem 27, the standard deviation σ_F is at most $\frac{1}{2}\sqrt{1/500} = \frac{1}{2}\sqrt{.002} = .02236$.

Let us decide on a 5 percent significance level. According to Appendix D, $p(|z| \leqslant 1.96) = .95$. Thus any normally distributed variable is within 1.96σ of μ with probability .95. Since f is approximately normally distributed with mean p, we shall reject any value of p if the observed value of f is not within 1.96σ of p. Since σ_F is at most .02236, we can use the larger number $(1.96)(.02236) = .0438$ instead of $1.96\sigma_F$.

In our case the observed value of f was $280/500 = .560$. We therefore reject all values of p outside the range $.560 \pm .0438$. Thus, the possible values of p are between .516 and .604. If p were outside this range, then the observed frequency $f = .56$ would be "too far" from p. We have thus found a 95 percent *confidence interval* for p: 51.6 percent $\leqslant p \leqslant$ 60.4 percent. We *do not say* that p is in this interval with 95 percent or more probability. The true value of p is fixed and is not a random variable. However, we do say that if p were not in this interval, an event occurred that had probability smaller than .05.

We can thus issue this statement to the press: "We claim that between 51.6 and 60.4 percent of the voters of this city favor candidate A. Our polling company is accurate more than 95 percent of the time."

The man in the street sees only $n = 500$ and a population of 1,000,000. Therefore, he feels that the sample is far too small. As a pollster, you see a *possibility* that $p < .516$ or $p > .604$, but you realize that if this were so, the results of your poll were a fluke that could occur only with probability smaller than .05. The man in the street should concentrate on the assumptions of independence and randomness, because these are the factors that are most difficult to achieve. In 1936 a magazine took a famous poll predicting Landon's victory over Roosevelt. They polled people at random from a telephone book, thereby eliminating a substantial (pro-Roosevelt) population that had no telephones. The mathematical techniques of this chapter, as well as much more refined and sophisticated techniques, are well known to pollsters. They know, however, that it is not as easy to obtain randomness and independence as it is to define these concepts.

EXERCISES

1. Prove, algebraically, that the largest value of $y = p - p^2$ is $y = \frac{1}{4}$ and that this occurs only at $p = \frac{1}{2}$. (*Hint*: Let $p = \frac{1}{2} + x$. Then find y as a function of x.)

2. Devise a test to decide, at the 5 percent significance level, whether a coin is fair or not. Do this for 10 tosses, 100 tosses, and 500 tosses.

3. As in Exercise 2, devise a test to decide at the 10 percent significance level whether a coin is fair. Use 100 tosses. Similarly, find a test at the 2 percent significance level.

4. Suppose a poll of 300 people shows that 180 people like brand X and 120 people detest brand X. Assuming that these 300 people were chosen independently and at random from a population of 10,000, find a 95 percent confidence interval for the percentage of people in the town who like brand X.

5. As in Exercise 4, find a 90 percent confidence interval and a 98 percent confidence interval.

6. The Weather Bureau in a certain city predicts rain with probabilities. (They might say that there is a 50 percent chance of rain.) Some people say that these are actual probabilities, but some say that the Weather Bureau is merely avoiding responsibility. It is therefore decided to check if these are accurate probabilities. For the next 50 predictions of "rain with probability 50 percent" it is observed whether it rained or not. Devise a test that might reject the assumption (at the 5 percent significance level) that the Weather

Bureau knows what it is talking about when it predicts rain with probability 50 percent.

7. An experiment consists in tossing 2 coins. How many experiments should be performed if the relative frequency of 2 heads is to be between .23 and .27 inclusive with probability 98 percent?

8. As in Exercise 7, how many experiments should be performed if the relative frequency of 2 heads is to be .24 or higher, with probability 95 percent?

9a. It is claimed that a coin has probability .2 or less of landing heads. You are willing to toss the coin 25 times. Devise a test to determine, at the 10 percent significance level, whether to reject the assumption $p \leqslant .2$.
b. Similarly, devise a test to reject the assumption $p \geqslant .2$.

10. One thousand people are sampled at a certain hour and it is found that 120 of them resented being polled. Find a 90 percent confidence interval for the percentage of people who resent being sampled at that hour. (Assume that the 1,000 people were selected independently and at random. This implies that 1 person *could* be sampled more than once – a dangerous procedure for people who are irate at being polled.)

11. A manufacturer sends a company a large shipment of nails. He claims that 20 percent of the nails are made of aluminum. The company president chose 150 nails at random and found only 20 aluminum nails. Granted that the president should have devised his experiment first, should he claim that he did not get the right amount of aluminum nails?

12. In Exercise 11 find a 95 percent confidence interval for the percentage of aluminum nails in the shipment. Use the method of Example 28.

*6 POISSON DISTRIBUTION

The Poisson distribution is a distribution which arises so often in practice that tables of its values have been extensively tabulated. Our approach is to treat it as the limiting value of a binomial distribution as $n \to \infty$ and $p \to 0$, where $np = \lambda$, some fixed number.

29 Theorem

Let λ be a fixed number and let $x \geqslant 0$ be an integer. Then

$$\lim_{n \to \infty} b\left(x; n, \frac{\lambda}{n}\right) = \lambda^x e^{-\lambda}/x! \tag{7.18}$$

Proof: Let $p = \lambda/n$. By definition,

$$b(x; n, p) = \binom{n}{x} p^x q^{n-x}$$

$$= \frac{n!}{x!(n-x)!} \frac{p^x}{q^x} q^n$$

$$= \frac{n!}{x!(n-x)!} \frac{\lambda^x}{n^x} \frac{1}{q^x} \left(1 - \frac{\lambda}{n}\right)^n$$

$$= \frac{\lambda^x}{x!} \left(1 - \frac{\lambda}{n}\right)^n \frac{n!}{(n-x)!n^x} \frac{1}{q^x} \qquad (7.19)$$

As $n \to \infty$, we have $[1 - (\lambda/n)^n \to e^{-\lambda}$, by Equation 4.6. Furthermore,

$$\frac{n!}{n^x(n-x)!} = \frac{n(n-1) \cdots (n-x+1)}{n^x} = \frac{n}{n} \frac{n-1}{n} \cdots \frac{n-x+1}{n}$$

$$= 1 \left(1 - \frac{1}{n}\right) \cdots \left(1 - \frac{x-1}{n}\right)$$

Since x is held fixed and each factor approaches 1 as $n \to \infty$, we have

$$\frac{n!}{(n-x)!n^x} \to 1 \qquad \text{as } n \to \infty \qquad (7.20)$$

Also, $q^x = [1 - (\lambda/n)]^x \to 1$ as $n \to \infty$, because x is held fixed. Hence, taking limits in Equation 7.19, we obtain

$$\lim b\left(x; n, \frac{\lambda}{n}\right) = \frac{\lambda^x}{x!} e^{-\lambda}(1)(1) = \frac{\lambda^x}{x!} e^{-\lambda}$$

This is the result.

30 Definition

The *Poisson distribution*, with parameter λ, is the distribution $p(x; \lambda)$ defined for $x = 0, 1, \ldots, n, \ldots$ by the formula

$$p(x; \lambda) = \frac{\lambda^x}{x!} e^{-\lambda} \qquad (7.21)$$

For small values of x, we have

$$p(0; \lambda) = e^{-\lambda}$$

$$p(1; \lambda) = \lambda e^{-\lambda}$$

$$p(2; \lambda) = \frac{\lambda^2}{2} e^{-\lambda}$$

Equation 7.18 merely states that $b(x; n, \lambda/n) \to p(x; \lambda)$ as $n \to \infty$. Thus we have

$$b(x; n, \lambda/n) \approx p(x; \lambda) \qquad \text{if } n \text{ is large} \qquad (7.22)$$

Equivalently, setting $p = \lambda/n$, we have

$$b(x; n, p) \approx p(x; \lambda) \qquad (\lambda = np) \qquad (7.23)$$

(If n is large, p is small, and $\lambda = np$ is moderately sized.)

The approximation 7.22 is seen to be useful if Bernoulli trials are repeated a large number of times, but the probability p of a success on each trial is small. In that case the probability of x successes is easily approximated in Equation 7.23 by using the Poisson distribution $p(x; \lambda)$ with $\lambda = np$. The values of $p(x; \lambda)$ are tabulated for various values of λ and x in Appendix F.

31 Example

The probability of winning a certain game is .01. a. If you play the game 150 times, what is the probability that you will win exactly twice? b. What is the probability that you will win 2 or more times?

Method. a. Here $p = .01$, $n = 150$. We assume that the games are independent. The required answer is $b(2; 150, .01)$. Here $\lambda = np = 150 \times .01 = 1.5$. By Equation 7.23 the Poisson approximation is

$$b(2; 150, .01) \approx p(2; 1.5)$$

Referring to Appendix F, under the column $\lambda = 1.5$ we find $p(2; 1.5) = .2510$. This is the required answer. [*Note:* Actually $p(2; 1.5) = [(1.5)^2/2]e^{-1.5}$. Appendix F merely evaluates this number.]

b. It is easier to compute the probability of 1 or 0 wins and take complements. Thus $b(1; 150, .01) + b(0; 150, .01) \approx p(1; 1.5) + p(0; 1.5) = .3347 + .2231 = .5578$. Thus the probability of 2 or more wins is approximately $1 - .5578 = .4422$. This completes the problem.

Note that we have assumed that $p(x; \lambda)$ is, for fixed λ, a distribution. This requires that

$$p(0; \lambda) + p(1; \lambda) + \cdots + p(n; \lambda) + \cdots = \sum_{n=0}^{\infty} p(n; \lambda) = 1 \qquad (7.24)$$

[An infinite sum is required because $p(x; \lambda)$ is defined for all nonnegative integers, not just finitely many.] But this equation is true, because by Equation 4.5 we have

$$p(0; \lambda) + p(1; \lambda) + \cdots + p(n; \lambda) + \cdots$$
$$= e^{-\lambda} + \lambda e^{-\lambda} + \frac{\lambda^2}{2!} e^{-\lambda} + \cdots + \frac{\lambda^n}{n!} e^{-\lambda} + \cdots$$
$$= e^{-\lambda} \left(1 + \lambda + \frac{\lambda^2}{2!} + \cdots + \frac{\lambda^n}{n!} + \cdots \right) = e^{-\lambda} \cdot e^{\lambda} = 1$$

Equation 7.24 was implicitly used above in part b. We wanted $p(2; 1.5) + p(3; 1.5) + \cdots$. This is $p(0; 1.5) + p(1; 1.5) + p(2; 1.5) + \cdots - p(0; 1.5) + p(1; 1.5) = 1 - [p(0; 1.5) + p(1; 1.5)]$. It was this number that we computed in part b.

Which would you rather play—a game with probability .01 for 150 times, or a game with probability .001 for 1,500 times? In each game the expected number of wins is 1.5. Theorem 29 shows that not only do the 2 have the same means, but they also have, approximately, the same distributions. In each case the probability of x wins is approximately $p(x; 1.5)$. The Poisson distribution is determined by 1 parameter, λ, unlike the binomial distribution, which needs $2 : n$ and p.

Since $\lambda = np$ is the *mean* of the binomial distribution, it is natural to say that the mean of the Poisson distribution is its parameter λ. Similarly, $\sigma = \sqrt{npq} = \sqrt{\lambda q} \approx \sqrt{\lambda}$ (because $q = 1 - (\lambda/n) \approx 1$ if n is large). Therefore, it is also natural to say that the standard deviation of the Poisson distribution is $\sqrt{\lambda}$. Both of these results may be proved directly using extensions of Equations 5.15 and 6.18 for infinite sums.

32   Example

An insurance company has insured 10,000 homes against fire. Assuming that the probability of fire in a house during a year is .0002 and that all fires are independent, find the probability that the company will have to pay off on 4 or more homes during a year.

Method.   Here $n = 10,000$, $p = .0002$, so $\lambda = np = 2$. The probability of 3 or less fires during a year is $p(0; 2) + p(1; 2) + p(2; 2) + p(3; 2) = .8571$. Hence the required probability is $1 - .8571 = .1429$.

One of the standard uses of the Poisson distribution is for continuous processes of a certain type. Example 33 gives one such application.

33   Example

A Geiger counter is placed near a radioactive substance in such a way that it registers an average of 10 counts per minute. (Each count occurs when an emitted particle hits the Geiger counter.) During a fixed minute, compute the probability that 7 or more counts occur.

Method.   Clearly we cannot proceed unless some physical and mathematical assumptions are made. (Ultimately these assumptions are tested by experiment.) It appears that particles are emitted randomly. Let us divide the minute into a large number n of equal parts. We shall assume that the process of emitting particles is a random procedure and that the probability of a particle hitting the Geiger counter in any one of these small periods of time is p. (p is small.) We also assume that hits on any 2 of these small time intervals are independent. We also assume that the time intervals are so small that 2 hits in one of these intervals cannot occur, or happen so infrequently that we can ignore the possibility. In that case we have approximated this process by a binomial distribution with parameters p and n. (A hit is regarded as a success.) The hypothesis is that $np = 10$, so the distribution of the number x of hits (total hits during a minute) is approximated by the Poisson distribution $p(x; 10)$. Here we expect a better approximation as $n \to \infty$, because presumably the possibility of 2 hits in an interval, if the time interval is divided more finely, will become more and more negligible as $n \to \infty$. It is therefore reasonable to regard $p(x; 10)$ as exactly equal to the distribution for the number of hits in 1 minute.

The required answer is therefore $p(7) + p(8) + \cdots$, where we have written $p(x; 10) = p(x)$. Since $p(0) + \cdots + p(6) = .1302$, from Appendix F, the required probability is $1 - .1302 = .8698$.

The method of this example applies to continuous processes in which a finite number of successes occur in a given time. As Example 33 shows, we require independence of success in different time intervals and a negligible probability of simultaneous successes. In these cases the probability of x successes in a given time interval is $p(x; \lambda)$, where λ is regarded as the average number of successes. For, as before, we may break up the given time intervals into a large number of pieces so that the probability of a success in each of these intervals is the same value p. Ignoring the possibility of 2 successes in any 1 of these intervals, or a success at the end points of these intervals, we have n Bernoulli trials with a small probability p of success on each trial. Letting $n \to \infty$ and assuming that $np = \lambda$ in all cases, we have a Poisson distribution in the limit. We shall not consider the precise conditions that a continuous process must satisfy to be a Poisson distribution.

Some examples where a Poisson distribution might apply (in the absence of other evidence) are the number of telephone calls in an office from 2 P.M. to 5 P.M. (λ = the average number of calls in this period), the number of accidents in the United States during the hours 12 noon through 12 midnight on a Saturday (here, a multiple accident, such as a pileup, must be counted as only 1 accident to avoid too many simultaneous "successes"), and the number of wrong notes played by a pianist during an hour's concert (although independence of the occurrence of wrong notes might depend upon the performer's experience and personality).

The reader is cautioned against assuming that any continuous random process that yields a finite number of successes is governed by a Poisson distribution. In the examples above we do not expect a Poisson distribution for the number of office calls if the office has only 1 telephone and a very talkative secretary. Can you explain why?

EXERCISES

Where appropriate, use the Poisson approximation to the binomial distribution.

1. Assume that 8 percent of the population is colorblind. If 50 people are chosen at random, what is the probability that at most 1 is colorblind?

2. Using Appendix F, sketch the graphs of $y = p(x; 2)$ and $y = p(x; 4)$.

3. A marksman has probability .01 of hitting the bulls-eye of a target. Assuming independence of trials, what is the probability that he hits the bulls-eye at least once in 50 shots? How many shots should he take so that he will hit the bulls-eye at least once, with probability .90 or more?

4. A manufacturing process produces a defective part with probability .001. A quality-control engineer will test 3,000 of these parts. He wants a number N such that the probability of N or more defectives is smaller than .05. Find N.

5. Assume that the probability of a twin birth is .01. During the next 250 deliveries at the hospital, what is the probability of 2 or more twin deliveries?

6. Prove: $p(x+1; \lambda) = (\lambda/x+1)p(x; \lambda)$.

7. For fixed λ, determine the value or values of x at which $p(x; \lambda)$ is as large as possible. (*Hint:* Use Exercise 6.)

8. Prove:
$$p(0; \lambda_1)p(0; \lambda_2) = p(0; \lambda_1+\lambda_2)$$
$$p(0, \lambda_1)p(1; \lambda_2) + p(0; \lambda_2)p(1; \lambda_1) = p(1; \lambda_1+\lambda_2)$$

and, in general,

$$\sum_{n=0}^{x} p(n; \lambda_1)p(x-n; \lambda_2) = p(x; \lambda_1+\lambda_2)$$

9. Algebra textbooks show that there can be no construction for the tri-section of an angle using straight edge and compass in certain specified ways. Despite this, every year a few "constructions" are published that prove incorrect in one detail or other.

    a. Give an argument in favor of the hypothesis that the number of such incorrect constructions per year has a Poisson distribution. Also give an argument against this hypothesis.

    b. Assume that the number of published faulty constructions has a Poisson distribution, with an average of 6 constructions per year. What is the probability that, next year, no angle-trisection constructions are published?

APPENDIXES

APPENDIX A
VALUES
OF
e^{-x}

x	e^{-x}	x	e^{-x}	x	e^{-x}	x	e^{-x}
0.00	1.0000	.45	.6376	.90	.4066	1.35	.2592
.01	0.9900	.46	.6313	.91	.4025	1.36	.2567
.02	.9802	.47	.6250	.92	.3985	1.37	.2541
.03	.9704	.48	.6188	.93	.3946	1.38	.2516
.04	.9608	.49	.6126	.94	.3906	1.39	.2491
.05	.9512	.50	.6065	.95	.3867	1.40	.2466
.06	.9418	.51	.6005	.96	.3829	1.41	.2441
.07	.9324	.52	.5945	.97	.3791	1.42	.2417
.08	.9231	.53	.5886	.98	.3753	1.43	.2393
.09	.9139	.54	.5827	.99	.3716	1.44	.2369
.10	.9048	.55	.5769	1.00	.3679	1.45	.2346
.11	.8958	.56	.5712	1.01	.3642	1.46	.2322
.12	.8869	.57	.5655	1.02	.3606	1.47	.2299
.13	.8781	.58	.5599	1.03	.3570	1.48	.2276
.14	.8694	.59	.5543	1.04	.3535	1.49	.2254
.15	.8607	.60	.5488	1.05	.3499	1.50	.2231
.16	.8521	.61	.5434	1.06	.3465	1.51	.2209
.17	.8437	.62	.5379	1.07	.3430	1.52	.2187
.18	.8353	.63	.5326	1.08	.3396	1.53	.2165
.19	.8270	.64	.5273	1.09	.3362	1.54	.2144
.20	.8187	.65	.5220	1.10	.3329	1.55	.2122
.21	.8106	.66	.5169	1.11	.3296	1.56	.2101
.22	.8025	.67	.5117	1.12	.3263	1.57	.2080
.23	.7945	.68	.5066	1.13	.3230	1.58	.2060
.24	.7866	.69	.5016	1.14	.3198	1.59	.2039
.25	.7788	.70	.4966	1.15	.3166	1.60	.2019
.26	.7711	.71	.4916	1.16	.3135	1.61	.1999
.27	.7634	.72	.4868	1.17	.3104	1.62	.1979
.28	.7558	.73	.4819	1.18	.3073	1.63	.1959
.29	.7483	.74	.4771	1.19	.3042	1.64	.1940
.30	.7408	.75	.4724	1.20	.3012	1.65	.1920
.31	.7334	.76	.4677	1.21	.2982	1.66	.1901
.32	.7261	.77	.4630	1.22	.2952	1.67	.1882
.33	.7189	.78	.4584	1.23	.2923	1.68	.1864
.34	.7118	.79	.4538	1.24	.2894	1.69	.1845
.35	.7047	.80	.4493	1.25	.2865	1.70	.1827
.36	.6977	.81	.4449	1.26	.2837	1.71	.1809
.37	.6907	.82	.4404	1.27	.2808	1.72	.1791
.38	.6839	.83	.4360	1.28	.2780	1.73	.1773
.39	.6771	.84	.4317	1.29	.2753	1.74	.1755
.40	.6703	.85	.4274	1.30	.2725	1.75	.1738
.41	.6637	.86	.4232	1.31	.2698	1.76	.1720
.42	.6570	.87	.4190	1.32	.2671	1.77	.1703
.43	.6505	.88	.4148	1.33	.2645	1.78	.1686
.44	.6440	.89	.4107	1.34	.2618	1.79	.1670

From DIFFERENTIAL AND INTEGRAL CALCULUS, Houghton Mifflin. Copyright ©, 1960, by James R. F. Kent, pages 470–473.

VALUES OF e^{-x}

x	e^{-x}	x	e^{-x}	x	e^{-x}	x	e^{-x}
1.80	.1653	2.25	.1054	3.45	.0317	6.40	.0017
1.81	.1637	2.26	.1044	3.50	.0302	6.50	.0015
1.82	.1620	2.27	.1033	3.55	.0287	6.60	.0014
1.83	.1604	2.28	.1023	3.60	.0273	6.70	.0012
1.84	.1588	2.29	.1013	3.65	.0260	6.80	.0011
1.85	.1572	2.30	.1003	3.70	.0247	6.90	.0010
1.86	.1557	2.31	.0993	3.75	.0235	7.00	.0009
1.87	.1541	2.32	.0983	3.80	.0224	8.00	.00034
1.88	.1526	2.33	.0973	3.85	.0213	9.00	.00012
1.89	.1511	2.34	.0963	3.90	.0202	10.00	.00005
1.90	.1496	2.35	.0954	3.95	.0193		
1.91	.1481	2.36	.0944	4.00	.0183		
1.92	.1466	2.37	.0935	4.05	.0174		
1.93	.1451	2.38	.0926	4.10	.0166		
1.94	.1437	2.39	.0916	4.15	.0158		
1.95	.1423	2.40	.0907	4.20	.0150		
1.96	.1409	2.41	.0898	4.25	.0143		
1.97	.1395	2.42	.0889	4.30	.0136		
1.98	.1381	2.43	.0880	4.35	.0129		
1.99	.1367	2.44	.0872	4.40	.0123		
2.00	.1353	2.45	.0863	4.45	.0117		
2.01	.1340	2.46	.0854	4.50	.0111		
2.02	.1327	2.47	.0846	4.55	.0106		
2.03	.1313	2.48	.0837	4.60	.0101		
2.04	.1300	2.49	.0829	4.65	.0096		
2.05	.1287	2.50	.0821	4.70	.0091		
2.06	.1275	2.55	.0781	4.75	.0087		
2.07	.1262	2.60	.0743	4.80	.0082		
2.08	.1249	2.65	.0707	4.85	.0078		
2.09	.1237			4.90	.0074		
2.10	.1225	2.70	.0672	4.95	.0071		
2.11	.1212	2.75	.0639	5.00	.0067		
2.12	.1200	2.80	.0608	5.10	.0061		
2.13	.1188	2.85	.0578	5.20	.0055		
2.14	.1177	2.90	.0550	5.30	.0050		
2.15	.1165	2.95	.0523	5.40	.0045		
2.16	.1153	3.00	.0498	5.50	.0041		
2.17	.1142	3.05	.0474	5.60	.0037		
2.18	.1130	3.10	.0450	5.70	.0033		
2.19	.1119	3.15	.0429	5.80	.0030		
2.20	.1108	3.20	.0408	5.90	.0027		
2.21	.1097	3.25	.0388	6.00	.0025		
2.22	.1086	3.30	.0369	6.10	.0022		
2.23	.1075	3.35	.0351	6.20	.0020		
2.24	.1065	3.40	.0334	6.30	.0018		

APPENDIX B
VALUES
OF
e^x

x	e^x	x	e^x	x	e^x	x	e^x
0.00	1.0000	.45	1.5683	.90	2.4596	1.35	3.8574
.01	1.0101	.46	1.5841	.91	2.4843	1.36	3.8962
.02	1.0202	.47	1.6000	.92	2.5093	1.37	3.9354
.03	1.0305	.48	1.6161	.93	2.5345	1.38	3.9749
.04	1.0408	.49	1.6323	.94	2.5600	1.39	4.0149
.05	1.0513	.50	1.6487	.95	2.5857	1.40	4.0552
.06	1.0618	.51	1.6653	.96	2.6117	1.41	4.0960
.07	1.0725	.52	1.6820	.97	2.6379	1.42	4.1371
.08	1.0833	.53	1.6989	.98	2.6645	1.43	4.1787
.09	1.0942	.54	1.7160	.99	2.6912	1.44	4.2207
.10	1.1052	.55	1.7333	1.00	2.7183	1.45	4.2631
.11	1.1163	.56	1.7507	1.01	2.7456	1.46	4.3060
.12	1.1275	.57	1.7683	1.02	2.7732	1.47	4.3492
.13	1.1388	.58	1.7860	1.03	2.8011	1.48	4.3929
.14	1.1503	.59	1.8040	1.04	2.8292	1.49	4.4371
.15	1.1618	.60	1.8221	1.05	2.8577	1.50	4.4817
.16	1.1735	.61	1.8404	1.06	2.8864	1.51	4.5267
.17	1.1853	.62	1.8589	1.07	2.9154	1.52	4.5722
.18	1.1972	.63	1.8776	1.08	2.9447	1.53	4.6182
.19	1.2092	.64	1.8965	1.09	2.9743	1.54	4.6646
.20	1.2214	.65	1.9155	1.10	3.0042	1.55	4.7115
.21	1.2337	.66	1.9348	1.11	3.0344	1.56	4.7588
.22	1.2461	.67	1.9542	1.12	3.0649	1.57	4.8066
.23	1.2586	.68	1.9739	1.13	3.0957	1.58	4.8550
.24	1.2712	.69	1.9937	1.14	3.1268	1.59	4.9037
.25	1.2840	.70	2.0138	1.15	3.1582	1.60	4.9530
.26	1.2969	.71	2.0340	1.16	3.1899	1.61	5.0028
.27	1.3100	.72	2.0544	1.17	3.2220	1.62	5.0531
.28	1.3231	.73	2.0751	1.18	3.2544	1.63	5.1039
.29	1.3364	.74	2.0959	1.19	3.2871	1.64	5.1552
.30	1.3499	.75	2.1170	1.20	3.3201	1.65	5.2070
.31	1.3634	.76	2.1383	1.21	3.3535	1.66	5.2593
.32	1.3771	.77	2.1598	1.22	3.3872	1.67	5.3122
.33	1.3910	.78	2.1815	1.23	3.4212	1.68	5.3656
.34	1.4049	.79	2.2034	1.24	3.4556	1.69	5.4195
.35	1.4191	.80	2.2255	1.25	3.4903	1.70	5.4739
.36	1.4333	.81	2.2479	1.26	3.5254	1.71	5.5290
.37	1.4477	.82	2.2705	1.27	3.5609	1.72	5.5845
.38	1.4623	.83	2.2933	1.28	3.5966	1.73	5.6407
.39	1.4770	.84	2.3164	1.29	3.6328	1.74	5.6973
.40	1.4918	.85	2.3396	1.30	3.6693	1.75	5.7546
.41	1.5068	.86	2.3632	1.31	3.7062	1.76	5.8124
.42	1.5220	.87	2.3869	1.32	3.7434	1.77	5.8709
.43	1.5373	.88	2.4109	1.33	3.7810	1.78	5.9299
.44	1.5527	.89	2.4351	1.34	3.8190	1.79	5.9895

Adapted from DIFFERENTIAL AND INTEGRAL CALCULUS, Houghton Mifflin. Copyright ©, 1960, by James R. F. Kent, pages 470–473.

VALUES OF e^x

x	e^x	x	e^x	x	e^x	x	e^x
1.80	6.0496	2.25	9.4877	3.45	31.500	5.70	298.87
1.81	6.1104	2.26	9.5831	3.50	33.115	5.75	314.19
1.82	6.1719	2.27	9.6794	3.55	34.813	5.80	330.30
1.83	6.2339	2.28	9.7767	3.60	36.598	5.85	347.23
1.84	6.2965	2.29	9.8749	3.65	38.475	5.90	365.04
1.85	6.3598	2.30	9.9742	3.70	40.447	5.95	383.75
1.86	6.4237	2.31	10.074	3.75	42.521	6.00	403.43
1.87	6.4883	2.32	10.176	3.80	44.701	6.05	424.11
1.88	6.5535	2.33	10.278	3.85	46.993	6.10	445.86
1.89	6.6194	2.34	10.381	3.90	49.402	6.15	468.72
1.90	6.6859	2.35	10.486	3.95	51.935	6.20	492.75
1.91	6.7531	2.36	10.591	4.00	54.598	6.25	518.01
1.92	6.8210	2.37	10.697	4.05	57.397	6.30	544.57
1.93	6.8895	2.38	10.805	4.10	60.340	6.35	572.49
1.94	6.9588	2.39	10.913	4.15	63.434	6.40	601.85
1.95	7.0287	2.40	11.023	4.20	66.686	6.45	632.70
1.96	7.0993	2.41	11.134	4.25	70.105	6.50	665.14
1.97	7.1707	2.42	11.246	4.30	73.700	6.55	699.24
1.98	7.2427	2.43	11.359	4.35	77.478	6.60	735.10
1.99	7.3155	2.44	11.473	4.40	81.451	6.65	772.78
2.00	7.3891	2.45	11.588	4.45	85.627	6.70	812.41
2.01	7.4633	2.46	11.705	4.50	90.017	6.75	854.06
2.02	7.5383	2.47	11.822	4.55	94.632	6.80	897.85
2.03	7.6141	2.48	11.941	4.60	99.484	6.85	943.88
2.04	7.6906	2.49	12.061	4.65	104.58	6.90	992.27
2.05	7.7679	2.50	12.182	4.70	109.95	6.95	1043.1
2.06	7.8460	2.55	12.807	4.75	115.58	7.00	1096.6
2.07	7.9248	2.60	13.464	4.80	121.51	8.00	2981.0
2.08	8.0045	2.65	14.154	4.85	127.74	9.00	8103.1
2.09	8.0849			4.90	134.29	10.00	22026
2.10	8.1662	2.70	14.880	4.95	141.17		
2.11	8.2482	2.75	15.643	5.00	148.41		
2.12	8.3311	2.80	16.445	5.05	156.02		
2.13	8.4149	2.85	17.288	5.10	164.02		
2.14	8.4994	2.90	18.174	5.15	172.43		
2.15	8.5849	2.95	19.106	5.20	181.27		
2.16	8.6711	3.00	20.086	5.25	190.57		
2.17	8.7583	3.05	21.115	5.30	200.34		
2.18	8.8463	3.10	22.198	5.35	210.61		
2.19	8.9352	3.15	23.336	5.40	221.41		
2.20	9.0250	3.20	24.533	5.45	232.76		
2.21	9.1157	3.25	25.790	5.50	244.69		
2.22	9.2073	3.30	27.113	5.55	257.24		
2.23	9.2999	3.35	28.503	5.60	270.43		
2.24	9.3933	3.40	29.964	5.65	284.29		

APPENDIX C
SQUARE ROOTS

n	$\sqrt{n}$	$\sqrt{10n}$	n	$\sqrt{n}$	$\sqrt{10n}$	n	$\sqrt{n}$	$\sqrt{10n}$
1	1.0000	3.1623	36	6.0000	18.974	71	8.4261	26.646
2	1.4142	4.4721	37	6.0828	19.235	72	8.4853	26.833
3	1.7321	5.4772	38	6.1644	19.494	73	8.5440	27.019
4	2.0000	6.3246	39	6.2450	19.748	74	8.6023	27.203
5	2.2361	7.0711	40	6.3246	20.000	75	8.6603	27.386
6	2.4495	7.7460	41	6.4031	20.248	76	8.7178	27.568
7	2.6458	8.3666	42	6.4807	20.494	77	8.7750	27.749
8	2.8284	8.9443	43	6.5574	20.736	78	8.8318	27.928
9	3.0000	9.4868	44	6.6332	20.976	79	8.8882	28.107
10	3.1623	10.000	45	6.7082	21.213	80	8.9443	28.284
11	3.3166	10.488	46	6.7823	21.448	81	9.0000	28.460
12	3.4641	10.954	47	6.8557	21.679	82	9.0554	28.636
13	3.6056	11.402	48	6.9282	21.909	83	9.1104	28.810
14	3.7417	11.832	49	7.0000	22.136	84	9.1652	28.983
15	3.8730	12.247	50	7.0711	22.361	85	9.2195	29.155
16	4.0000	12.649	51	7.1414	22.583	86	9.2736	29.326
17	4.1231	13.038	52	7.2111	22.804	87	9.3274	29.496
18	4.2426	13.416	53	7.2801	23.022	88	9.3808	29.665
19	4.3589	13.784	54	7.3485	23.238	89	9.4340	29.833
20	4.4721	14.142	55	7.4162	23.452	90	9.4868	30.000
21	4.5826	14.491	56	7.4833	23.664	91	9.5394	30.166
22	4.6904	14.832	57	7.5498	23.875	92	9.5917	30.332
23	4.7958	15.166	58	7.6158	24.083	93	9.6437	30.496
24	4.8990	15.492	59	7.6811	24.290	94	9.6954	30.659
25	5.0000	15.811	60	7.7460	24.495	95	9.7468	30.822
26	5.0990	16.125	61	7.8102	24.698	96	9.7980	30.984
27	5.1962	16.432	62	7.8740	24.900	97	9.8489	31.145
28	5.2915	16.733	63	7.9373	25.100	98	9.8995	31.305
29	5.3852	17.029	64	8.0000	25.298	99	9.9499	31.464
30	5.4772	17.321	65	8.0623	25.495			
31	5.5678	17.607	66	8.1240	25.690			
32	5.6569	17.889	67	8.1854	25.884			
33	5.7446	18.166	68	8.2462	26.077			
34	5.8310	18.439	69	8.3066	26.268			
35	5.9161	18.708	70	8.3666	26.458			

Reprinted from ELEMENTS OF FINITE PROBABILITY by J. L. Hodges and E. Lehmann, by permission of the publisher, Holden-Day, Inc., page 219.

AREAS UNDER THE NORMAL CURVE

The table entry is $N(z)$, the area under the standard normal curve from 0 to z. The area from z to ∞ is $.5000 - N(z)$. The area from $-z$ to z is $2N(z)$.

$N(z)$

z	.00	.01	.02	.03	.04	.05	.06	.07	.08	.09
0.0	.0000	.0040	.0080	.0120	.0160	.0199	.0239	.0279	.0319	.0359
0.1	.0398	.0438	.0478	.0517	.0557	.0596	.0636	.0675	.0714	.0753
0.2	.0793	.0832	.0871	.0910	.0948	.0987	.1026	.1064	.1103	.1141
0.3	.1179	.1217	.1255	.1293	.1331	.1368	.1406	.1443	.1480	.1517
0.4	.1554	.1591	.1628	.1664	.1700	.1736	.1772	.1808	.1844	.1879
0.5	.1915	.1950	.1985	.2019	.2054	.2088	.2123	.2157	.2190	.2224
0.6	.2257	.2291	.2324	.2357	.2389	.2422	.2454	.2486	.2517	.2549
0.7	.2580	.2611	.2642	.2673	.2704	.2734	.2764	.2794	.2823	.2852
0.8	.2881	.2910	.2939	.2967	.2995	.3023	.3051	.3078	.3106	.3133
0.9	.3159	.3186	.3212	.3238	.3264	.3289	.3315	.3340	.3365	.3389
1.0	.3413	.3438	.3461	.3485	.3508	.3531	.3554	.3577	.3599	.3621
1.1	.3643	.3665	.3686	.3708	.3729	.3749	.3770	.3790	.3810	.3830
1.2	.3849	.3869	.3888	.3907	.3925	.3944	.3962	.3980	.3997	.4015
1.3	.4032	.4049	.4066	.4082	.4099	.4115	.4131	.4147	.4162	.4177
1.4	.4192	.4207	.4222	.4236	.4251	.4265	.4279	.4292	.4306	.4319
1.5	.4332	.4345	.4357	.4370	.4382	.4394	.4406	.4418	.4429	.4441
1.6	.4452	.4463	.4474	.4484	.4495	.4505	.4515	.4525	.4535	.4545
1.7	.4554	.4564	.4573	.4582	.4591	.4599	.4608	.4616	.4625	.4633
1.8	.4641	.4649	.4656	.4664	.4671	.4678	.4686	.4693	.4699	.4706
1.9	.4713	.4719	.4726	.4732	.4738	.4744	.4750	.4756	.4761	.4767
2.0	.4772	.4778	.4783	.4788	.4793	.4798	.4803	.4808	.4812	.4817
2.1	.4821	.4826	.4830	.4834	.4838	.4842	.4846	.4850	.4854	.4857
2.2	.4861	.4864	.4868	.4871	.4875	.4878	.4881	.4884	.4887	.4890
2.3	.4893	.4896	.4898	.4901	.4904	.4906	.4909	.4911	.4913	.4916
2.4	.4918	.4920	.4922	.4925	.4927	.4929	.4931	.4932	.4934	.4936
2.5	.4938	.4940	.4941	.4943	.4945	.4946	.4948	.4949	.4951	.4952
2.6	.4953	.4955	.4956	.4957	.4959	.4960	.4961	.4962	.4963	.4964
2.7	.4965	.4966	.4967	.4968	.4969	.4970	.4971	.4972	.4973	.4974
2.8	.4974	.4975	.4976	.4977	.4977	.4978	.4979	.4979	.4980	.4981
2.9	.4981	.4982	.4982	.4983	.4984	.4984	.4985	.4985	.4986	.4986
3.0	.4987	.4987	.4987	.4988	.4988	.4989	.4989	.4989	.4990	.4990

z	0.674	1.282	1.645	1.960	2.326	2.576	3.090	3.291	3.891
$N(z)$	.25	.40	.45	.475	.49	.495	.499	.4995	.49995
$2N(z)$	.50	.80	.90	.95	.98	.99	.998	.999	.9999

From Mosteller, Rourke and Thomas, PROBABILITY WITH STATISTICAL APPLICATIONS, 2nd ed., 1970, Addison-Wesley, Reading, Mass., page 473.

APPENDIX E
BINOMIAL DISTRIBUTION

$$b(x; n, p) = \binom{n}{x} p^x q^{n-x}$$

n	x	p .1	.2	.3	.4	.5
2	0	.8100	.6400	.4900	.3600	.2500
	1	.1800	.3200	.4200	.4800	.5000
	2	.0100	.0400	.0900	.1600	.2500
3	0	.7290	.5120	.3430	.2160	.1250
	1	.2430	.3840	.4410	.4320	.3750
	2	.0270	.0960	.1890	.2880	.3750
	3	.0010	.0080	.0270	.0640	.1250
4	0	.6561	.4096	.2401	.1296	.0625
	1	.2916	.4096	.4116	.3456	.2500
	2	.0486	.1536	.2646	.3456	.3750
	3	.0036	.0256	.0756	.1536	.2500
	4	.0001	.0016	.0081	.0256	.0625
5	0	.5905	.3277	.1681	.0778	.0312
	1	.3280	.4096	.3602	.2592	.1562
	2	.0729	.2048	.3087	.3456	.3125
	3	.0081	.0512	.1323	.2304	.3125
	4	.0004	.0064	.0284	.0768	.1562
	5	.0000	.0003	.0024	.0102	.0312
6	0	.5314	.2621	.1176	.0467	.0156
	1	.3543	.3932	.3025	.1866	.0938
	2	.0984	.2458	.3241	.3110	.2344
	3	.0146	.0819	.1852	.2765	.3125
	4	.0012	.0154	.0595	.1382	.2344
	5	.0001	.0015	.0102	.0369	.0938
	6	.0000	.0001	.0007	.0041	.0156
7	0	.4783	.2097	.0824	.0280	.0078
	1	.3720	.3670	.2471	.1306	.0547
	2	.1240	.2753	.3177	.2613	.1641
	3	.0230	.1147	.2269	.2903	.2734
	4	.0026	.0287	.0972	.1935	.2734
	5	.0002	.0043	.0250	.0774	.1641
	6	.0000	.0004	.0036	.0172	.0547
	7	.0000	.0000	.0002	.0016	.0078
8	0	.4305	.1678	.0576	.0168	.0039
	1	.3826	.3355	.1977	.0896	.0312
	2	.1488	.2936	.2965	.2090	.1094
	3	.0331	.1468	.2541	.2787	.2188
	4	.0046	.0459	.1361	.2322	.2734
	5	.0004	.0092	.0467	.1239	.2188
	6	.0000	.0011	.0100	.0413	.1094
	7	.0000	.0001	.0012	.0079	.0312
	8	.0000	.0000	.0001	.0007	.0039

From Brunk, AN INTRODUCTION TO MATHEMATICAL STATISTICS, 1960, Ginn & Co., pages 363–365.

$$b(x; n, p) = \binom{n}{x} p^x q^{n-x}$$

n	x	.1	.2	.3	.4	.5
				p		
9	0	.3874	.1342	.0404	.0101	.0020
	1	.3874	.3020	.1556	.0605	.0176
	2	.1722	.3020	.2668	.1612	.0703
	3	.0446	.1762	.2668	.2508	.1641
	4	.0074	.0661	.1715	.2508	.2461
	5	.0008	.0165	.0735	.1672	.2461
	6	.0001	.0028	.0210	.0743	.1641
	7	.0000	.0003	.0039	.0212	.0703
	8	.0000	.0000	.0004	.0035	.0176
	9	.0000	.0000	.0000	.0003	.0020
10	0	.3487	.1074	.0282	.0060	.0010
	1	.3874	.2684	.1211	.0403	.0098
	2	.1937	.3020	.2335	.1209	.0439
	3	.0574	.2013	.2668	.2150	.1172
	4	.0112	.0881	.2001	.2508	.2051
	5	.0015	.0264	.1029	.2007	.2461
	6	.0001	.0055	.0368	.1115	.2051
	7	.0000	.0008	.0090	.0425	.1172
	8	.0000	.0001	.0014	.0106	.0439
	9	.0000	.0000	.0001	.0016	.0098
	10	.0000	.0000	.0000	.0001	.0010
11	0	.3138	.0859	.0198	.0036	.0005
	1	.3835	.2362	.0932	.0266	.0054
	2	.2131	.2953	.1998	.0887	.0269
	3	.0710	.2215	.2568	.1774	.0806
	4	.0158	.1107	.2201	.2365	.1611
	5	.0025	.0388	.1321	.2207	.2256
	6	.0003	.0097	.0566	.1471	.2256
	7	.0000	.0017	.0173	.0701	.1611
	8	.0000	.0002	.0037	.0234	.0806
	9	.0000	.0000	.0005	.0052	.0269
	10	.0000	.0000	.0000	.0007	.0054
	11	.0000	.0000	.0000	.0000	.0005
12	0	.2824	.0687	.0138	.0022	.0002
	1	.3766	.2062	.0712	.0174	.0029
	2	.2301	.2835	.1678	.0639	.0161
	3	.0852	.2362	.2397	.1419	.0537
	4	.0213	.1329	.2311	.2128	.1208
	5	.0038	.0532	.1585	.2270	.1934
	6	.0005	.0155	.0792	.1766	.2256
	7	.0000	.0033	.0291	.1009	.1934
	8	.0000	.0005	.0078	.0420	.1208
	9	.0000	.0001	.0015	.0125	.0537
	10	.0000	.0000	.0002	.0025	.0161
	11	.0000	.0000	.0000	.0003	.0029
	12	.0000	.0000	.0000	.0000	.0002

		p				
n	x	.1	.2	.3	.4	.5
13	0	.2542	.0550	.0097	.0013	.0001
	1	.3672	.1787	.0540	.0113	.0016
	2	.2448	.2680	.1388	.0453	.0095
	3	.0997	.2457	.2181	.1107	.0349
	4	.0277	.1535	.2337	.1845	.0873
	5	.0055	.0691	.1803	.2214	.1571
	6	.0008	.0230	.1030	.1968	.2095
	7	.0001	.0058	.0442	.1312	.2095
	8	.0000	.0011	.0142	.0656	.1571
	9	.0000	.0001	.0034	.0243	.0873
	10	.0000	.0000	.0006	.0065	.0349
	11	.0000	.0000	.0001	.0012	.0095
	12	.0000	.0000	.0000	.0001	.0016
	13	.0000	.0000	.0000	.0000	.0001
14	0	.2288	.0440	.0068	.0008	.0001
	1	.3559	.1539	.0407	.0073	.0009
	2	.2570	.2501	.1134	.0317	.0056
	3	.1142	.2501	.1943	.0845	.0222
	4	.0349	.1720	.2290	.1549	.0611
	5	.0078	.0860	.1963	.2066	.1222
	6	.0013	.0322	.1262	.2066	.1833
	7	.0002	.0092	.0618	.1574	.2095
	8	.0000	.0020	.0232	.0918	.1833
	9	.0000	.0003	.0066	.0408	.1222
	10	.0000	.0000	.0014	.0136	.0611
	11	.0000	.0000	.0002	.0033	.0222
	12	.0000	.0000	.0000	.0005	.0056
	13	.0000	.0000	.0000	.0001	.0009
	14	.0000	.0000	.0000	.0000	.0001
15	0	.2059	.0352	.0047	.0005	.0000
	1	.3432	.1319	.0305	.0047	.0005
	2	.2669	.2309	.0916	.0219	.0032
	3	.1285	.2501	.1700	.0634	.0139
	4	.0428	.1876	.2186	.1268	.0417
	5	.0105	.1032	.2061	.1859	.0916
	6	.0019	.0430	.1472	.2066	.1527
	7	.0003	.0138	.0811	.1771	.1964
	8	.0000	.0035	.0348	.1181	.1964
	9	.0000	.0007	.0116	.0612	.1527
	10	.0000	.0001	.0030	.0245	.0916
	11	.0000	.0000	.0006	.0074	.0417
	12	.0000	.0000	.0001	.0016	.0139
	13	.0000	.0000	.0000	.0003	.0032
	14	.0000	.0000	.0000	.0000	.0005
	15	.0000	.0000	.0000	.0000	.0000

APPENDIX F
POISSON
DISTRIBUTION

$$p(x;\lambda) = \lambda^x e^{-x}/x!$$

x \ λ	.1	.2	.3	.4	.5	.6	.7	.8	.9	1.0	x
0	.9048	.8187	.7408	.6703	.6065	.5488	.4966	.4493	.4066	.3679	0
1	.0905	.1637	.2222	.2681	.3033	.3293	.3476	.3595	.3659	.3679	1
2	.0045	.0164	.0333	.0536	.0758	.0988	.1217	.1438	.1647	.1839	2
3	.0002	.0011	.0033	.0072	.0126	.0198	.0284	.0383	.0494	.0613	3
4		.0001	.0003	.0007	.0016	.0030	.0050	.0077	.0111	.0153	4
5				.0001	.0002	.0004	.0007	.0012	.0020	.0031	5
6							.0001	.0002	.0003	.0005	6
7										.0001	7

x \ λ	1.5	2.0	2.5	3.0	3.5	4.0	4.5	5.0	6.0	7.0	8.0	9.0	10	x
0	.2231	.1353	.0821	.0498	.0302	.0183	.0111	.0067	.0025	.0009	.0003	.0001	.0000	0
1	.3347	.2707	.2052	.1494	.1057	.0733	.0500	.0337	.0149	.0064	.0027	.0011	.0005	1
2	.2510	.2707	.2565	.2240	.1850	.1465	.1125	.0842	.0446	.0223	.0107	.0050	.0023	2
3	.1255	.1804	.2138	.2240	.2158	.1954	.1687	.1404	.0892	.0521	.0286	.0150	.0076	3
4	.0471	.0902	.1336	.1680	.1888	.1954	.1898	.1755	.1339	.0912	.0573	.0337	.0189	4
5	.0141	.0361	.0668	.1008	.1322	.1563	.1708	.1755	.1606	.1277	.0916	.0607	.0378	5
6	.0035	.0120	.0278	.0504	.0771	.1042	.1281	.1462	.1606	.1490	.1221	.0911	.0631	6
7	.0008	.0034	.0099	.0216	.0385	.0595	.0824	.1044	.1377	.1490	.1396	.1171	.0901	7
8	.0001	.0009	.0031	.0081	.0169	.0298	.0463	.0653	.1033	.1304	.1396	.1318	.1126	8
9		.0002	.0009	.0027	.0066	.0132	.0232	.0363	.0688	.1014	.1241	.1318	.1251	9
10			.0002	.0008	.0023	.0053	.0104	.0181	.0413	.0710	.0993	.1186	.1251	10
11				.0002	.0007	.0019	.0043	.0082	.0225	.0452	.0722	.0970	.1137	11
12				.0001	.0002	.0006	.0016	.0034	.0113	.0264	.0481	.0728	.0948	12
13					.0001	.0002	.0006	.0013	.0052	.0142	.0296	.0504	.0729	13
14						.0001	.0002	.0005	.0022	.0071	.0169	.0324	.0521	14
15							.0001	.0002	.0009	.0033	.0090	.0194	.0347	15
16									.0003	.0014	.0045	.0109	.0217	16
17									.0001	.0006	.0021	.0058	.0128	17
18										.0002	.0009	.0029	.0071	18
19										.0001	.0004	.0014	.0037	19
20											.0002	.0006	.0019	20
21											.0001	.0003	.0009	21
22												.0001	.0004	22
23													.0002	23
24													.0001	24

Based on Brunk, AN INTRODUCTION TO MATHEMATICAL STATISTICS, 1960, Ginn & Co., pages 221, 371–374.

Answers to Selected Odd-Numbered Exercises

CHAPTER 1

Section 1

1. $f_A = 119/600 = .198$; $f_B = 203/600 = .338$; $f_C = 278/600 = .463$.

3. $1945: f_M = 1,467/2,858 = .513$, $f_F = 1,391/2,858 = .487$; $1946: f_M = .514$, $f_F = .486$; 1947, 1948, 1949: $f_M = .513$, $f_F = .487$. The large samples involved indicate that it is somewhat more likely that a male is born than a female.

5a. $s_1, s_2, s_3, s_4, s_5, s_6$, where s_i is the outcome "high die is i."

7a. $H_0, H_1, \ldots, H_{10}$, where H_i is the outcome "exactly i heads are tossed."

9a. There are infinitely many possible outcomes: $H_1, H_2, H_3, \ldots$, and H_∞. H_i is the outcome "the first head occurs on the ith trial," and H_∞ is the event "a head never turns up," which is a theoretical possibility.

Section 2

1i. Not a probability space since $.1 + .2 + .3 + .4 + .5 \neq 1.0$.

  ii. A probability space.

  iii. A probability space.

3. $1/5$.

5. $p(A) = \frac{1}{5}$ and $p(B) = p(C) = \frac{2}{5}$.

7i. No, since statistical probabilities are limits of f_i as N tends to infinity.

  ii. Yes, because the $p(s_i)$ satisfy the conditions 1.4 and 1.5.

9. $1/39$.

Section 3

1. $A_1 = \emptyset$, $A_2 = \{X\}$, $A_3 = \{Y\}$, $A_4 = \{Z\}$, $A_5 = \{X, Y\}$, $A_6 = \{X, Z\}$, $A_7 = \{Y, Z\}$, and $A_8 = \{X, Y, Z\} = S$.

3. The event is $A = \{2, 4, 6\}$. $p(A) = .624$. The relative frequency of A in the first 100 trials was .61. The relative frequency in the second 100 trials was .63.

5. The event is $A = \{7, 11\}$. $p(A) = .223$. The relative frequency of A was .244. $p(\{6, 7, 8\}) = .445$.

7. The event is $\{9, 19, 29, 39, 49, 59, 69, 79, 89,$ and 90 through 99$\}$. The probability is $19/100 = .19$.

9a. $3/7$

  b. $1/3$

  c. $4/21$

  d. $4/21$

  e. $2/21$

11a. $.09$

  b. $.33$

  c. $.20$

  d. $.17$

    e. .10

13a. $n(A) = 3. p(A) = .8$

    b. .5

    c. $\sum_{s \in \{x,y,z\}} p(s)$

Section 4

1. $\frac{11}{36} = .306$

1st \ 2nd	1	2	3	4	5	6
1				$(1,4)$		
2				$(2,4)$		
3				$(3,4)$		
4	$(4,1)$	$(4,2)$	$(4,3)$	$(4,4)$	$(4,5)$	$(4,6)$
5				$(5,4)$		
6				$(6,4)$		

3. $20/36 = .556$

5a. $16/32 = .5$

    b. $8/32 = .25$

7. $11/32 = .344$

9. The sample points are HHH, HHT, HTH, THH, HTT, THT, TTH, and TTT. The probabilities are $\frac{1}{8}, \frac{3}{8}, \frac{3}{8}$, and $\frac{1}{8}$ respectively.

11a. $1/3$

    b. $2/5$

    c. $1/3$

    d. $3/10$

    e. $1/6$

13. The probabilities are respectively 1/36, 10/36, and 25/36. To three decimal places, these are .028, .278, and .694.

Section 5

1. $p(\text{Manhattan}) = 1,698/7,782 = .218$

    $p(\text{Bronx}) \quad\;\; = 1,425/7,782 = .183$

    $p(\text{Brooklyn}) \;\; = 2,627/7,782 = .338$

    $p(\text{Queens}) \quad\; = 1,810/7,782 = .233$

    $p(\text{Richmond}) = \quad 222/7,782 = .029$

3. There are 25 prime numbers between 1 and 100 inclusive.

CHAPTER 2

Section 1

1. $p = 6/16 = .375$

	HH	HT	TH	HH
HH	HHHH	HHHT	HHTH	HHTT
HT	HTHH	HTHT	HTTH	HTTT
TH	THHH	THHT	THTH	THTT
TT	TTHH	TTHT	TTTH	TTTT

3. The probability that both are green is $42/341 = .123$. The probability that both have the same color is $162/341 = .475$.

5. n(three letter words) $= 26^3 = 17,576$. n(three letter words of the type consonant-vowel-consonant) $= 21 \cdot 5 \cdot 21 = 2,205$.

7. Let $A = \{N, S, E, W\}$ (the points of a compass), and let $D = \{F, L, R\}$ (forward, left, and right). Then, Paths $= A \times D \times D \times D \times D$. n(Paths) $= 4(3^4) = 324$. The man will be unable to stroll on a different path every day for a year. First path = ELRRF. Second path = WFRFR.

9. There are $3^3 = 27$ different ways of answering, and there are 30 students. Therefore, some tests will be identical.

11a. $7^4 = 2,401$

b. $7^3 = 343$

13. It is possible to produce the same amounts in different ways. For example, $3\cent = 1\cent + 2\cent$. This was not possible in Exercise 12.

Section 2

1. $3/7 = .429$

3. $20 \cdot 19 \cdot 18/23 \cdot 22 \cdot 21 = .644$

5. $12 \cdot 11 \cdot 10 \cdot 9/12 \cdot 12 \cdot 12 \cdot 12 = .573$

7. $6 \cdot 5 \cdot 4 \cdot 3 \cdot 2/6^5 = 5/54 = .093$

9a. $4/22,100 = .0002$

b. $44/22,100 = .002$

c. $1,096/22,100 = .050$

d. $720/22,100 = .033$

e. $52/22,100 = .002$

f. $3,744/22,100 = .169$

g. $16,440/22,100 = .744$

11. $p_0 + p_1 + p_2 + p_3 = 1$, and $p_1 = p_2$, $p_0 = p_3 = 2/17$. Therefore, $1 = 2p_0 + 2p_2 = 4/17 + 2p_2$. Solving for p_2, we find $p_2 = 13/34$.

13a. $1,260$

b. 780

c. 660

d. $1,100$

e. 60
f. 114

Section 3

1a. 720
  b. 336
  c. 20
  d. 117,600
  e. 220
  f. 74/81
  g. 230,300
  h. 5
  i. 720
  j. 70
3a. 12/11!
  b. 17/10!7!
  c. 19!/11!8!
5. 495
7. 171
9a. Population = the 26 letters; size is indeterminate; ordered sample with replacement.
  b. Population = {0, 1, 2, 3, 4, 5, 6, 7, 8, 9}; size 4; ordered sample with replacement.
  c. Population = 52 cards; size 5; can be ordered or unordered without replacement.
  d. Population = {1, 2, 3, 4, 5, 6}; size 8; ordered or unordered with replacement.
  e. Population = {H, T}; size 10; ordered or unordered with replacement.
  f. Population = all students in the graduating class; size 5; ordered sample without replacement.
  g. Population = all baseball players; size 9; ordered or unordered without replacement.
  h. Population = all American League teams; size = number of teams in the American League; ordered sample without replacement.
  i. Population = all National league teams; size = number of places in the first division; unordered sample without replacement.

11. The number of hands is $2{,}598{,}960 = \binom{52}{5}$. The probability is $\binom{13}{5} / \binom{52}{5} = .0005$.

13. 30! /10!

Section 4

1a. $\binom{14}{6}$

b. $\binom{13}{7}$

c. $\binom{n+1}{r}$

d. $\binom{x+3}{r+4}$

9a. $1 + 100x + 4{,}950x^2$

b. $x^{50} - 50x^{49}y + 1{,}225x^{48}y^2$

c. $1 - 70t + 2{,}415t^2$

d. $p^n + np^{n-1}q + \dfrac{n(n-1)}{2}p^{n-2}q^2$

11a. .737

b. 1.010

c. 1.000

d. .968

13. $\binom{24}{3}$

15. 36; 28

Section 5

5a. $1/1{,}820 = .0005$

b. $24/455 = .053$

c. $64/455 = .141$

7. 1,260

9. $\binom{k+n-1}{k-1}$

11. Even is more likely.

13a. $\binom{80}{7} \Big/ \binom{100}{7}$

b. $\binom{80}{5}\binom{20}{2} \Big/ \binom{100}{7}$

c. $\left[\binom{80}{5}\binom{20}{2} + \binom{80}{6}\binom{20}{1} + \binom{80}{7}\right] \Big/ \binom{100}{7}$

CHAPTER 3

Section 1

1a. $\{1, 2, 3, 5, 7, 9\}$
  b. $\{1, 4, 6, 8, 9, 10\}$
  c. $\{1, 9\}$
  d. $\{3, 5, 7\}$
  e. $\{1, 9\}$
  f. $\emptyset$
  g. $\{1, 2, 3, 4, 5, 6, 7, 8, 9\}$
  h. $\{10\}$
3a. $\bar{E}$
  b. $P \cap E$
  c. $\bar{E} \cap Sq$
  d. $T \cap E$
  e. $Sq \cup P$
  f. not possible
  g. $\bar{P} \cap Sq$
  h. not possible
5a. $\bar{A} = \{d, e, f, g\}$;
    $\bar{B} = \{b, d, g\}$;
    $\bar{C} = \{a, b\}$;
    $A \cup B = \{a, b, c, e, f\}$;
    $B \cap C = \{c, e, f\}$;
    $A \cap B \cap C = \{c\}$;
    $\bar{A} \cup B = \{a, c, d, e, f, g\}$
  b. $p(\bar{A}) = .65$;
    $p(\bar{B}) = .50$;
    $p(\bar{C}) = .30$;
    $p(A \cup B) = .70$;
    $p(B \cap C) = .40$;
    $p(A \cap B \cap C) = .05$;
    $p(\bar{A} \cup B) = .80$
7. 12 percent
9. p(at least one 6 among 3 dice) $= 91/216 = .421$;
    p(at least one 6 among 4 dice) $= 671/1{,}296 = .518$
11. 0.4
13. 280
15a. $750/1{,}326 = .566$
  b. $8{,}852/22{,}100 = .401$
17a. $.2 \leqslant p(A \cup B) \leqslant .3$
  b. $.5 \leqslant p(A \cup B) \leqslant .9$
  c. $.7 \leqslant p(A \cup B) \leqslant 1$

d. $.9 \leqslant p(A \cup B) \leqslant 1$
e. $.95 \leqslant p(A \cup B) \leqslant 1$
19a. $0 \leqslant p(A \cap B) \leqslant .1$
b. $0 \leqslant p(A \cap B) \leqslant .4$
c. $.3 \leqslant p(A \cap B) \leqslant .6$
d. $.1 \leqslant p(A \cap B) \leqslant .2$
e. $.85 \leqslant p(A \cap B) \leqslant .9$

Section 2

1. 30.8 percent
3. $.167/.667 = .250$
5. $p(A|B) = p(A)/p(B) \geqslant p(A)$ since $p(B) \leqslant 1$ and $p(B)$ is positive. Equality holds if and only if $p(A) = 0$ or $p(B) = 1$ (A is impossible or B is certain).

7. $\binom{4}{2}\binom{48}{3}\bigg/\left[\binom{52}{5} - \binom{48}{5}\right]$

9a. $p(\text{all aces}) = 1/5{,}525 = .0002$
b. $p(\text{all aces}|\text{different suits}) = 1/2{,}197 = .0005$
11. $9/230 = .039$
13. If $p(B) = .9$ and $p(A) = .6$, then $5/9 \leqslant p(A|B) \leqslant 6/9$, or $p(A|B)$ is between .555 and .667. If $p(B) = .99$, then $59/99 \leqslant p(A|B) \leqslant 60/99$, or $p(A|B)$ is between .584 and .594.

Section 3

1. 3.84 percent
3. $p(\text{faulty stamp}) = .07$;
$p(\text{dealer } X|\text{faulty stamp}) = 4/7 = .571$
5. $24/59 = .407$
7. $4/13 = .308$
9. $p(\text{2-headed}|\text{three consecutive heads}) = 8/17 = .471$;
$p(\text{head on 4th trial}|\text{three consecutive heads}) = 25/34 = .735$
11. $9/19 = .473$
13. $208/2{,}210 = .094$

Section 4

1. $p(\text{red card}) = 26/52 = \frac{1}{2}$; $p(\text{ace}) = 4/52 = 1/13$; $p(\text{red ace}) = 2/52 = 1/26$. Since $(1/2)(1/13) = 1/26$, the events are independent.
3. $p(\text{6 on first die}) = 1/6$; $p(\text{sum is 8}) = 5/36$; $p(\text{sum is 8 and a 6 is on first die}) = 1/36$. Since $(1/6)(5/36) \neq 1/36$, the events are not independent.
5. $n(A)n(B) = n(S)n(A \cap B)$
9a. $5/144 = .035$
b. $113/144 = .785$

The assumption is that the events are independent.

11. $1 - (1 - 1/365)^{10}$, or approximately $10/365$

13. $1 - (.9)^{20}$

15. $p(A_1 \cup A_2 \cup A_3) = p_1 + p_2 + p_3 - (p_1 p_2 + p_1 p_3 + p_2 p_3) + p_1 p_2 p_3$
$= p(A_1) + p(A_2) + p(A_3) - [p(A_1 \cap A_2) + p(A_1 \cap A_3) + p(A_2 \cap A_3)]$
$+ p(A_1 \cap A_2 \cap A_3)$

17. 192

21. $p(2 \text{ before } 3) = \frac{1}{2}[1 - (\frac{2}{3})^{10}] = .491$
$p(\text{no 2 or 3}) = (\frac{2}{3})^{10} = .017$

Section 5

9. No. It is necessary to know the value of $p(B_1)/p(B)$. $\frac{1}{3} \leqslant p(A|B) \leqslant \frac{1}{2}$. If $p(B_1) = p(B_2)$, then $p(A|B) = 5/12$. If $p(B_1) = 2p(B_2)$, then $p(A|B) = 4/9$.

11.

1st \ 2nd	1	2	3	4	5	6
1	(1, 1)	(1, 2)	(1, 3)	(1, 4)	(1, 5)	(1, 6)
2	(2, 1)	(2, 2)	(2, 3)	(2, 4)	(2, 5)	(2, 6)
3	(3, 1)	(3, 2)	(3, 3)	(3, 4)	(3, 5)	(3, 6)
4	(4, 1)	(4, 2)	(4, 3)	(4, 4)	(4, 5)	(4, 6)
5	(5, 1)	(5, 2)	(5, 3)	(5, 4)	(5, 5)	(5, 6)
6	(6, 1)	(6, 2)	(6, 3)	(6, 4)	(6, 5)	(6, 6)

$p(\text{sum} = 7) = .18$;
$p(\text{sum} = 7 \text{ or } 11) = .26$;
$p(\text{high die is 6}) = .36$;
$p(\text{either a 6 or a 1 appears}) = .64$

15. $p(s_2, t_1) = .20$
$p(s_2, t_3) = .12$
$p(s_3, t_1) = .25$
$p(s_3, t_2) = .10$
$p(s_3, t_3) = .15$

CHAPTER 4

Section 1

1. $.299 < P_{100} < .366$

3. $n \geqslant 26$ times.

5. $.4288$

7. For $n = 4, .1042$. For $n = 25, .4831$

9. $.7135$

11. $.8118 < P_{10} < .8642$

15a. 1.11

  b. $.905$

  c. 2.718

  d. $.368$

Section 2

1. The probability that A hits the target first is

$$\frac{p}{p + p' - pp'}.$$

The probability that B hits the target first is

$$\frac{p'(1-p)}{p + p' - pp'}.$$

3. A has the better chance of hitting the target first. His probability is 9/17.

5. $p = .50$: probability $= .6$

  $p = .49$ at \$1 per game: probability $= .58$

  $p = .49$ at \$.10 per game: probability $= .36$

  $p = .51$ at \$1 per game: probability $= .62$

  $p = .51$ at \$.10 per game: probability $= .80$

Section 3

1. $.581$

3. $.794$

5. $3,168$

7. $2,200$

11a. $.3679$

  b. $.0613$

  c. $.6321$

  d. 1.0000

Section 4

1. 19/27 in both cases. A possible symmetry in the first case is:

1	2	3	4	5	6
↓	↓	↓	↓	↓	↓
1	4	3	2	6	5

A possible symmetry in the second case is

1	2	3	4	5	6
↓	↓	↓	↓	↓	↓
5	2	3	4	1	6

3. .088. A possible symmetry is

1	2	3	4	5	6
↓	↓	↓	↓	↓	↓
1	3	5	2	4	6

5. $739/1,632 = .453$

7. $176/1,275 = .138$

9. .1617

11. The probabilities of landing at Y and Z are equal.

CHAPTER 5

Section 1

1a. 10.0

b. 64

c. 1.4

d. 4.1

e. 3.5

f. 12.6

5. $\displaystyle\sum_{i=1}^{n} x_i^2 - 2a \sum_{i=1}^{n} x_i + na^2$

Section 2

1. $p(1) = .2; p(2) = .6; p(3) = .2$

3. $p(2) = p(12) = \frac{1}{36} = .028;$

$p(3) = p(11) = \frac{1}{18} = .056;$

$p(4) = p(10) = \frac{1}{12} = .083;$

$p(5) = p(9) = \frac{1}{9} = .111;$

$p(6) = p(8) = \frac{5}{36} = .139;$

$p(7) = \frac{1}{6} = .167$

5. $p(0) = p(5) = \frac{1}{32} = .031;$
$p(1) = p(4) = \frac{5}{32} = .156;$
$p(2) = p(3) = \frac{10}{32} = .313$

7. $p(-1) = \frac{37}{38} = .974; p(35) = \frac{1}{38} = .026$

9. $p(0) = \frac{1}{2} = .500$
$p(1) = \frac{2}{7} = .286$
$p(2) = \frac{1}{7} = .143$
$p(3) = \frac{2}{35} = .057$
$p(4) = \frac{1}{70} = .014$

Section 3

1. Average temperature $= 57.5°$. The median is $56.5°$.

3. Average value of $x = 3.5$. Average value of $x^2 = \frac{91}{6} = 15.17$.

7. Average number of aces is $\frac{2}{13} = .154$. The average number of diamonds is $\frac{1}{2} = .5$.

9. The expected high die is $\frac{161}{36} = 4.47$. The expected low die is $\frac{91}{36} = 2.53$.

11. 2.5

15. 2.7

17. $\frac{25}{16} = 1.56$

Section 4

1. 15

3a. 3.5

b. 7

c. 35

5. $\frac{20}{3} = 6.67$

7. 101

9. $E(X-1) = 0; E(X-1)^2 = 1; E(2X+1)^2 = 13$

Section 5

1. $8,280

3. $2,401/1,296 = 1.85$

5. 14. For k heads, the average number of tosses is

$$k + \sum_{i=1}^{k} i2^{k-1} = 2^{k+1} - 2.$$

Section 6

1.

x \ y	0	1	$p(x)$
1	.1	.2	.3
2	.4	0	.4
3	0	.3	.3
$q(y)$	.5	.5	

X and Y are not independent since $p(x, y) \neq p(x)q(y)$.

3. No.

5.

x \ y	1	2
1	.1	0
2	0	.1
3	.1	0
4	0	.1
5	.1	0
6	0	.1
7	.1	0
8	0	.1
9	.1	0
10	0	.1

Section 7

1. The value is $9/13 = .69$. A's correct strategy is $11/13H + 2/13T$. This is also B's correct strategy.

3. The value is 3. A plays T, and B plays H.

5. The value is $17/8 = 2.125$. A's correct strategy is to put out one finger with probability $\frac{7}{8}$, and two fingers with probability $\frac{1}{8}$. This is also B's correct strategy.

Section 8

5. 4.5

CHAPTER 6

Section 1

1. $\mu = .5, \sigma = .5$

3. $\mu = p, \sigma = p(1-p) = pq$

5. $m = 75.9, s = 13.7$

7. $E(X^3) - 3\mu\sigma^2 - \mu^3$

11. $\mu = 4.96, m = 4.99; \sigma = 1.15, s = 1.20$

Section 2

1. $p \geq \frac{15}{16} = .9375$

3. Change the extreme temperatures to $61.6°$ and $88.4°$.

5. $99 < x < 101$

7. probability $\geq .75$

Section 3

1. $\mu = 7, \sigma = 2.42$

3. $\mu = 80, \sigma = 4$

5. $\mu = 40, \sigma = 10.7$

7. 6.3

11. Expected winnings $= \$200$, standard deviation $= \$110$.

Section 4

1. $E(\bar{X}) = \frac{1}{4} = .25; \sigma(\bar{X}) = \sqrt{3}/40 = .043$

5. 25 experiments; 40 experiments

11a. $\mu = 1, \sigma = \sqrt{2}/2 = .707$

  b. $\mu = 1, \sigma = \sqrt{5}/10 = .224$

Section 5

1. Any $N_0 \geq pq/d^2f$

3. $N = 200,000$

CHAPTER 7

Section 1

1a. .15
 b. .575
 c. .175
 d. .175
5. $\frac{5}{12} = .417$

Section 2

1a. .0166
 b. .8849
 c. .0479
 d. .8414
3. $9.670 \leqslant x \leqslant 10.330$ with probability .9; $9.744 \leqslant x \leqslant 10.256$ with probability .8.
5a. 16.2
 b. 61.04% passing grade; 77.28% for an A.
7. $q = 82$; $q = 98$ for rare meetings.

Section 3

3. .7361
5. Expected number $= 210$, standard deviation $= 12.12$
7. .9298

Section 4

1. .7286 for 100 tosses;
 .8612 for 200 tosses;
 .9774 for 500 tosses
3. .8230; $13 \leqslant n \leqslant 27$ for probability larger than .90,
 $12 \leqslant n \leqslant 28$ for probability larger than .95,
 $9 \leqslant n \leqslant 31$ for probability larger than .99
5. .8790
7. .2483 = probability that 10 or more favor B,
 .9452 = probability that 15 or fewer favor A
9. .0516; $N = 17$
11. 40% or higher with probability .0951; 50% or higher with probability .0003.

Section 5

3. At the 10% significance level, reject the hypothesis that the coin is fair if the number of heads is not between 42 and 58 inclusive. At the 2%

significance level, reject the hypothesis if the number of heads is not between 38 and 62 inclusive.

5. The 90% confidence interval is $55.25\% \leq p \leq 64.75\%$.
   The 98% confidence interval is $53.28\% \leq p \leq 66.72\%$.

7. about 2,536

9a. Reject hypothesis if 9 or more heads turn up.

 b. Reject hypothesis if 1 or no heads turn up.

11. Since the number of aluminum nails should be 9 or less away from the mean 20 with probability .9476, an event with very small probability (.0524) has occurred. The president could claim a minor miracle, or he could check his 20% claim.

Section 6

1. .0916
3. .3935; about 250 shots
5. .7127
9b. .0025

INDEX

INDEX

Aristotle, 15
Average value, *see* Expectation

Bayes' Theorem, 100
Bernoulli's Theorem, 233
Binomial coefficients, 64
Binomial distribution, 252
Binomial theorem, 63
Births, by sex, 9
Bridge, 78

Chebyshev bounds, 220
Chebyshev's Theorem, 219
Coin tossing, 25
Combinations, 55
Complement of sets, 82
Conditional expectation, 180
Conditional probability, 92
Conditional relative frequency, 91

D'Alembert, 27, 29
Dependent events, 104
Dice, high number thrown, 2, 3
    sum, 9, 23
    tossed till 6 occurs, 7, 8

Distribution, 159
    binomial, 252
    normal, 244
    Poisson, 270
Division principle, 51

Elementary event, 12
Events, dependent, 104
    in a probability experiment, 17
    in a probability space, 18
    independent, 104–105
    mutually exclusive, 83
    probability of, 18
    relative frequency of, 17
Expectation, 168
    conditional, 180
    units of, 177
Expected value, *see* Expectation

Finite sets as sample space, 32
Frequency, 4

Games, theory of, 197

Histograms, 161

Inclusion-exclusion principle, 140
Independence of events, 104
    of random variables, 188
Independent trials, 226
Intersection of sets, 82

Joint distribution, 186

Law of averages, 6
Law of Large Numbers, 232
Limit, defined, 129

Mean value, *see* Expectation
Median, 171
Miracles, 7
Multinomial Theorem, 72
Multiplication principle, 45–46

Normal distribution, 244

Partitions, 114

Pascal's triangle, 62
Permutations, 54
Poisson distribution, 270
Poker, 76
Polling, 27, 254
Population, 52
Probability, conditional, 92
    definition, 12
    of an event, 18
    statistical definition, 5
Probability density, 237
       See also Distribution
Probability experiment, 3
Probability space, definition, 12
    infinite discrete, 33
    uniform, 14
Product probability space, 115
Product rule, 97
Product set, 38–39

Random variables, definition, 158
    distribution of, 159
    expectation of, 168
    independence of, 188
    product of, 175
    standard deviation of, 211
    sum of, 175
    variance of, 208
Random walk, 132–135, 137
Relative frequency, 4
    conditional, 91
    of an event, 18

Sample mean, 213
Sample point, 12
Sample space, 12
       See also Probability space
Sample standard deviation, 213
Sample variance, 213
Samples, classified, 52–53
    number of unordered, 67
Significance level, 263
Standard deviation, 211
    units of, 211
Symmetry, definition, 147
    effect on random variables, 201

Tree diagrams, 98
Triple plays, 218

Uniform probability space, 14
    probability in, 19
Union of sets, 82
Urns, 28

Variance, 208
    units of, 210